# Ethnobotanical Analysis of Wild Food Plants Traditionally Collected and Consumed

# Ethnobotanical Analysis of Wild Food Plants Traditionally Collected and Consumed

Edited by:

**Akansha Singh and Mohammad Waseemul Islam**

Delve
PUBLISHING

www.delvepublishing.com

**Ethnobotanical Analysis of Wild Food Plants Traditionally Collected and Consumed**

*Akansha Singh and Mohammad Waseemul Islam*

**Delve Publishing**
224 Shoreacres Road
Burlington, ON L7L 2H2
Canada
www.delvepublishing.com

Email: orders@arclereducation.com



ISBN: 978-1-77407-803-7 (Hardcover)

# ABOUT THE EDITORS

**Dr. Akansha Singh,** is presently working as Associate Professor & Head, Department of Genetics and Plant Breeding, College of Agriculture, Parul University, India. She graduated from the Banaras Hindu University, with doctorate in Genetics and Plant breeding in the year 2012. She has also worked as Research Scientist in Bioved Research Society, Allahabad, India. She was working as Postdoctoral fellow in University of Mumbai before joining her current position. She has authored a number of national and international publications.

**Dr. Waseem's** professional expertise lies in teaching and pre-clinical research on indigenous medicinal plants. He is the Former Director of Complementary Medicine Unit, Chairman of Ethical Committee and Head of the Research Group, at Ajman University, UAE. Dr. Waseem has worked for phytochemical, pharmacological and nutritional values of medicinal plants, is actively engaged in the preparation of the tentative plan of nutrition courses for Ajman University, and presented the importance of plant nutrition for sustainable development for global health. Dr. Waseem has over 100 research publications and presentations on phytoconstituents, pharmacology & toxicology of medicinal plants & herbal products. His professional record demonstrates enriched association with renowned educational institutions, namely, College of Medicine, University of Erbil, Iraq; College of Pharmacy, King Saud University, Riyadh, Saudi Arabia; Health Authority, Abu Dhabi, UAE; Dubai Medical College for Girls, Dubai, UAE and Ajman University, College of Pharmacy and Health Sciences, Ajman and UAE. Dr. Waseem has been actively involved in executing several scientific projects, co-authored the Encyclopedia of Medicinal Plants of UAE, published by Zayed Complex for Herbal Research and Traditional Medicine, printed in Al-Omera Printing and Publishing Co., Abu Dhabi, UAE. He was a reviewer for the International Journals and served as Chairman and Co-chairman at the conferences of phytomedicines and food derived from plants, selected by World Health Organization (WHO) as an advisor for Eastern Mediterranean Region in 2003. He has previously served as Head of the Department, supervising research projects & managing research team in Zayed Complex for Herbal Research and Traditional Medicine, Abu Dhabi, UAE. He has carried out research on medicinal, aromatic and poisonous plants in the College of Pharmacy, King Saud University, Riyadh, Saudi Arabia. He also taught and carried out several research projects in Dubai Medical College, affiliated with the Central Biomedical Research Laboratories as Executive Consultant & Research Manager. He was awarded a plaque of appreciation for his significant contribution to the success of WHO Consultation on Quality Control of Herbal Medicine, Abu Dhabi, UAE and a shield from General Authority for Health Services for the Emirate of Abu Dhabi, UAE. He has been honored in recognition of his devotion, sincerity and his contributions in the fields of research and other academic activities from the College of Pharmacy, King Saud University, Riyadh. A commemorative ceremony to celebrate the scientific achievements of Dr. Waseem was held in the academic year 2013-14 on his excellence in research contribution, at College of Pharmacy and Health Sciences, Ajman University, UAE organized by the College of Pharmacy & Health Sciences.

# TABLE OF CONTENTS

# LIST OF FIGURES

# PREFACE

Wild plants are increasingly being used for domestic consumption in many countries around the world. Some of these species are multi-purpose and belong to multiple commodity categories. For instance, five Uapaca (Euphorbiaceae) species occurring in Malawi are commonly used to provide food, wine, herbal medicine, fuelwood and timber. Due to the versatility of this plant, suggestions for domestication, selection and improvement have been proposed.

The impact of wild plants on family subsistence use depends mostly on individual circumstances. With regards to food, although wildly picked plants may serve the entire community during a famine, no family relies entirely on wild resources throughout the year. Often, wild plants constitute a complex and mixed lifestyle, where their contributions come from different sources such as agricultural and wild -collected products (such as food, fiber, honey, beverages, medicinal plants and oils).

Wild plants can help meet the minimum living needs, or can be used as luxury goods; if an excessive amount is collected, it can be used in barter trade, sold in the local market or traded as a small business. For instance, in Turkey, a variety of wild plant foods (such as wild mushrooms) are gathered for personal and family consumption, and excess food is usually sold in a street or urban markets, typically by women in so-called "women's markets". Most wild food plants are consumed raw, mostly collected from young plants. Some examples include the stems of Rhubarb plant and *Rumex alpinus*. Other plants are consumed after some cooking, boiled in water or fried with oil and eggs. Examples include; *Borago officinalis* and *Portulaca oleracea*. There are also some wild tea, spices and condiment plants that must be dried before use.

The sustenance value of wild plants varies from one region to another. Although the main function of these plants is the supplementation of food obtained through household gardens and other agricultural forms, many cultivated or wild-harvested species also provide essential vitamins, seasonings, and other important compounds obtained from secondary products through the metabolism of alkaloids, phenols and essential oils.

For example, in Mediterranean countries, salads and leafy vegetables are often gathered from the wild to increase dietary diversity and nutrition. Meanwhile, in Crete, Greece, around 200 wild plant species are consumed, whether at home or in the restaurant, as a common dish or "mountain greens" specialty respectively. They are gathered in the wild by local farmers, who not only eat them themselves but also carry them to the market to increase their income. In Sicily, Italy, about 24 wild cruciferous plants of 16 genera are sometimes picked as vegetables.

This volume discusses wild edible plants in detail. Chapter 1 is an: introduction to wild edible plants, chapter 2: nutrient and anti-nutrient content of wild food plants, chapter 3: classification of wild edible plants, chapter 4: edible aquatic plants, chapter 5: wild edible plants found in forests, chapter 6: identifying common poisonous wild plants, chapter 7: domestication of wild food plants, chapter 8: ultimate foraging guide for edible wild plants, chapter 9: factors affecting wild edible plant and chapter 10: conservation of wild edible plants.

# 1

# *Introduction to Wild Edible Plants*

## CONTENTS

## 1.1 INTRODUCTION

**Figure 1.1:** An image of edible food plants growing wild.

Source: https://containergardening.wordpress.com/tag/balcony-garden-web/

There are more than 20,000 types of palatable edible plants on the planet, yet less than 20 species currently source of 90% of our food. Be that as it may, there are many less notable consumable plants from all around the globethat are both scrumptious and contain an immense nutritional value to the human body. In the life of human beings as early as history takes us, plant edibility has been a key survival factor in the lives of humans. The early human being used to live in the wild and there was an adaptation due to evolution as they were able to identify plants and animals that they could eat. Insects and bugs also apply here. Their edibility is something that was a key element to look at due to the possibility of poisoning or death from the plants. There are plant parts that cannot be eaten but other parts of the same plant can be eaten. This information was gathered and developed from the hunting and gathering habits of the early man. Collecting of berries, leaves, bark, roots and other parts of plants for the consumption use, medicinal uses, and other uses. This made it easy to identify the edible and non-edible plants for the human diet. This also offered insight into what could be given as foodto the domesticated animals as well (Angier, 2001).

The interaction with the plants even today in modern times including the five-sense interaction is a key factor in determining whether eat or not to eat the plant parts. The chapter provide detail insight about the plants that are edible and their subsequent parts. This information is especially useful to those that are in the wild and it can be very critical when it comes to surviving in the wild in the event of an unfortunate event such as a plane crash, a shipwreck, among many other accidents that could leave survivors. It can also be as simple as scouts getting lost or hikers getting lost on an adventure. Having your baring in check on what to eat and what not to eat in the wild could prove detrimental to those that live and those that die (Angier, 2001).

## 1.2 WHAT IS EDIBLE?

Majority of the time, where people stuck at remote wild location and they feel hunger, they are get attracted to colorful plants and their fruits and assumed them as edible However, these unknown plants can be poisionous such as Blue elderberry plant that can contains poisionous substance in their berries, leaves etc and when people consume them, it could be detrimental to their health. The assumption of how edible a plant is can be very dangerous and many at times ends in deadly results. Edible wild food plants are defined as those that have or contain one part or more parts that can be consumed as food if they are gathered only if they are gathered at the correct stages of their growth in addition to them being prepared in the right way. In saying one or more parts, it is common for plants to have different parts that can be edible. This could be the stems, the roots, the fruits, the leaves, seeds as well as the shoots. When a plant is defined as edible, this means that the plant has at least one part that can be eaten from the whole plant. For example, all but the cooked underground tuber of a potato plant is considered poisonous and have serious toxins. The potato itself has toxins but the preparation method and also the peeling process helps in the reduction of the toxins. This also applies to the Blue Elderberry plant because the flowers and the fully ripe berries are actually edible. This just directs individuals to have basic knowledge of wild plants hence focusing on the edible plants only of plants.

**Figure 1.2:** The image of Blue Elderberry plant growing in wild.

*Source:* https://gardenerspath.com/plants/fruit/best-elderberry-varieties/

When it comes to the collection methods, the plant parts need to be collected at the appropriate growth stage which is suitable for eating. Information on the correct stage of growth of the plant parts not only gives the best food, it also keeps you safe and healthy. Some plant parts such as ***Asclepias syriaca*** also known as the common milkweed has pods that contain seeds, which in the young stage are white in color, and at this stage, they can be eaten as cooked vegetable. After this stage, they turn brown indicating maturity and they gain toxins and becomes non-edible for human.

**Figure 1.3:** An image of a mature common milkweed seed pod.

*Source:https://marcpastorek.com/2015/02/15/planting-milkweed-seed-for-monarch-butterflies-nows-the-time/*

The aspect of proper preparation, some of the wild foods may not be edible unless they are processed in one way or another. Some of the ways may include peeling off the top layers of different parts, physical removal or detachment from other parts, leaching out parts of the plant parts with water -soluble substances, heating the plant parts to a certain temperature among many other ways that the wild foods can be processed. To be safe, it is important to be able to process the different parts of plants that are considered edible in the right way in order to avoid poisoning. The knowledge and information on the processing methods of different parts of plants that are edible but require processing is fundamental and critical to the individual's safety. A key point to note, the wild plants that are considered edible could be ingested into the body for the purposes of food or for medical use. There are a lot of plant parts that are used for medicinal reasons, but the same risks apply in the picking of the part at the right stage, the preparation method and also identifying what plant parts are safe and viable for medical use (Auerbach, 2011).

## 1.3 WHAT PLANT PARTS MORPH INTO FOOD

There are sure tendencies that plants have as they create. The more you think about these inclinations, the better you will have the option to make decisions about precisely at the point when a palatable part will be at its best. You will likewise have a superior thought of what may have turned out badly it something (taste or surface) doesn't go true to form. As a rule, vegetative plant parts develop from being delicate to getting stringy as they develop. ''Fast development'' also, ''end size'' are significant ideas in this section. These ideas will give you the most ideal approach to decide when a plant part will be ideal for eating. Each plant part will have its own account of development and size, so let's spread them individually. Any plant part experiences development until it arrives at its end size. Quick development results from ideal developing conditions, bringing about bigger stems, roots, leaves, and so forth. Development doesn't occur consistently over all plant parts. For example, a lower stem may have hardened into a strong non-developing accessory structure, and yet, it very well may be transporting supplements and water to developing parts of a plant. That lower developed stem is thin or even woody, while the developing new developments are delicate. Here we analyze the functionsfunctions that development and end size have on different portions of plants that is critical for flavor and surface. (Tull, 2013).

### 1.3.1 End Size

End size can allude to the entire plant once it quits developing. Each plant parts develop at their own rate and at their own time. The end size of each part is dictated by both hereditary qualities and how nature has sustained the plant. The end size of a leaf, for example, can be huge in the event that it has high levels of water reaching it, sufficient daylight exposure, and nutrition.

For most plant parts, synthetic and basic changes happen once the end size is reached. Those parts change from development to help and support structures. Fiber works to reinforce the leaf, and a portion of the material is needed for healthy development of the leaf.

### 1.3.2 Leaves

As a general rule, leaves like other plant parts, begin delicate and remain sensibly so until they arrive at their end size. Done be tricked by little leaf size, because a leaf is little doesn't mean it is youthful, delicate, or scrumptious. As simply stated, a little leaf may be little in light of the fact that it has been famished of water and supplements, arriving at its end size from the get-go throughout everyday life. This is frequently the situation in summer when the soil has been hard and dry for some time. On the other hand, you discover little leaves on an all-around watered plant, almost certainly, those leaves are youthful and as yet developing. Progressively develop plants can have quickly developing parts that are incredible for eating. A few leaves like those found on excavators' lettuce and wild spinach never get "as well" stringy to eat. These leaves develop in such fantastic conditions that even at eighteen weeks they have not yet arrived at their full size, despite being massive in size. They look delightful when found in the farmers' market. They were as yet delicate, and their harshness had not yet had an opportunity to get unfortunately unforgiving. In some normally unpleasant leaves, the quality of their sharpness is identified with moderate development and end size. Plants like dandelions or wild lettuce collect sharpness after some time. The quicker a leaf is developing, the more spread out and less exceptional that sharpness is. The slower the leaf is developing, or in the event that it has arrived at its end size, the more sharpness that is concentrated from exposure to the sun. Consider it along these lines: the more extended the leaves are exposed in the sun, the rough they become, and the harder it is to oversee that roughness in the kitchen. Along these lines, quickly developing leaves are not just increasingly delicate, they will in general be less harsh in those plants that produce severe flavors (Tull, 2013).

### 1.3.3 Stems

Stems give structure and backing to leaves, flowers, wild products, and seeds. Stems develop to extend the length of a plant, permitting new leaves and different structures to spread out. Contingent upon the plane, stems can develop upstanding, along the ground, under the ground, or they can spread outward. A few plants have evident stems, and some don't. A few plants by and large called biennials just develop stems in their second year of development. Stems can be an extraordinary wellspring of food on the off chance that you realize the ideal chance to utilize them in the plant's advancement. Much the same as leaves, stems that are effectively developing have increasingly delicate parts. It is quite often the tips of stems that bend effectively developing. Like leaves, stem parts that are not developing can be trimmed off to stimulate faster plant growth. The no-more developing regions will turn out to be progressively thin. The more drawn out the stem gets; the more fiber is set down to help the weight. Let's talk about the expected edibility of various phases of stem advancement in plants with palatable stems.

### 1.3.4 Shoots (Developing Tips of Plants)

Starting at the soil level when a plant is developing starting from the earliest stage. Some herbaceous plants like asparagus develop quickly in the spring from perpetual roots. They send up thick shoots that seem as though the asparagus we purchase from the basic food item. Things being what they are, asparagus is a youthful stem that hasn't panned out yet. At the point when it has everything, has that asparagus-twig look, it is quickly developing and will be sufficiently delicate to eat. It is likewise realized that there is a point close to the base of that stem that is too thin to even consider eating. So, strip it or on the other hand, snap it off to be disposed of. That flexible part is there to strenghten the stem so it can more readily support a taller plant. Asparagus that grows in such conditions is commonly picked when it is around ten inches in length. Quickly developing asparagus can be eaten as long as the fruit can be effortlessly snapped from the lengthening stem. There have been cases where people snap the delicate upper parts of stems up to two feet tall. After trimming, the developing branchlet tips are sufficiently delicate to eat when they are as yet maturing. Asparagus turns out to be as well thin to eat once the stem is just strong. Asparagus is regular on the grounds that a great many people know this plant or possibly what it resembles in the market or on a supper plate. Be that as it may, this equivalent rule applies to numerous wild foods (Tull, 2013).

**Figure 1.4:** An image of asparagus. As long as the top part of the plant appears to be tender and feels tender, the plant is harvestable and can be cooked to eat.

*Source:* https://www.thespruce.com/how-to-plant-and-grow-asparagus-1402814

## 1.4 POTENTIAL OF WILD FOODS

### 1.4.1 Why Eat Wild Foods?

There are numerous motivations to find out about, discover, assemble, also, eat wild foods. The essential reasons approaching are that it's fun, it fulfills an assortment of human needs, and this food is bravo profoundly, mentally and genuinely. Engaging with wild foods takes advantage of your daring soul, fulfills the drive to be progressively independent, what's more, satisfies the longing to communicate all the more personally with the earth and its abundance. It shows you a way to be an understudy of the plant world, a specialist, an onlooker, a history criminologist. You wind up spending increasingly "soil time" learning the mystery life of plants. Wild foods give a fun, information based recreational action that definitively draws in pass on heart and brain. The social affair of obtaining food legitimately from the earth, rather than finding a disinfected plastic enveloped item by the product segment of a grocery store, greatly affects the spirit. It carries us closer to the earth and to our far-off progenitors, and it incites sentiments of certainty and confidence. Realizing enough to scrounge, collect, and eat your own wild food is a basic experience.

The examination and utilization of wild foods can everything being equal become a deep-rooted open- air recreational, instructive, and culinary movement new fun, new food, and new interest for a lifetime (Tardío, J., Pardo-de-Santayana, M., & Morales, R., 2006).

### 1.4.2 Wild Foods are Good for Your Family

The entirety of the advantages picked up by you as an individual are intensified in a family setting. In the event that you are a parent who appreciates wild foods, it has the additional advantage of being extremely a good time for kids at any age. Getting to know one another as a family, investigating nature close-up while exploring, gathering what's more, cooking the plants will be times you and your children always remember. Including kids in the entire procedure from recognizing and gathering to getting ready and eating, will connect with their interest and start a long-lasting family movement. Children today are detached from food frameworks. In a significant manner, wild foods will train them where food comes from and there's the additional advantage of activity. Adolescence and grown-up obseity rates are a scourge in North America. In today's computer game, TV, and PC centered stationary presence, this open-air movement can help to give the sorts of physical exercise and dietary commitments that the parents and kids need to encourage a more advantageous way of life.

### 1.4.3 Rewarding Nature Deficit Disorder

Kids today are uninformed of precisely what food is, the place food originates from, what's more, how it gets into the structures we find in the general store. Adding to the issue is their seclusion from nature itself done investing a lot of energy in it in view of a steady flood from building up the dread of everything to everybody.

### 1.4.4 Wild Foods are Good for the Earth

Wild foods are a phenomenal scene for inspiring individuals to turn out to be better guardians of the earth. The investigation and dynamic utilization of wild foods carries you closer to nature in an amazingly personal manner since you are picking up food straightforwardly from the earth. You see all land not simply farmland and your nursery as a possible wellspring of food. So, any sort of contamination turns into all considerably more obvious. Second, palatable weeds are a must in privately developed maintainable nourishments.

### 1.4.5 Wild Foods are Good Nutritionally

Numerous individuals accept that wild foods merit seeking after in view of their colossal dietary substance. What's more, there is no doubt that wild foods speak to a brilliant wellspring of supplements. Without a doubt, many wild food proponents accept that they are definitive nourishment. It is immovably accepted that it can do only useful for North Americans to eat all the newer verdant green leaves and vegetables. That isn't to state you can't choose a solid eating routine from the new greens and vegetables as of now accessible in the grocery store in reality, you can. Individuals of every single dietary reasonableness will profit more from culinary and dietary assortment offered by wild foods. Omnivores will profit by including all the newer vegetation.

### 1.4.6 Wild Foods are Good for Your Wallet

In the event that you are living on a constrained or fixed pay, new grocery store vegetables can remove a genuine lump from your check. The utilization of effectively open palatable weeds can save your cash for different things and grow the assortment of your eating regimen. Wild foods can be a fence against expansion, what's more, fluctuating food costs, another option or expansion to some locally acquired nourishments, and an improvement in the nature of regular day to day existence.

### 1.4.7 Wild Foods are Good for Farmers and the Economy

Indeed, are weeds useful for farmers? Farmers, especially small scalefarmers, are scarcely scratching by nowadays. Many are losing their farms since they can't contend with the mega corporate homesteads developing ordinary yields. Small scalefarmers could begin having some expertise in more specialty crops like a portion of the wild foods provided in this book. They could likewise profit by using gleaners to harvest wild foods.

## 1.5 THE NUTRITION OF WILD FOODS

The ordinary classifications of supplements, we as a whole know are proteins, fats, sugars, nutrients, and minerals. Inside those classifications are specific supplements including some amino acids, fats, vitamin C, and so forth., that we need for our bodies to work. New substances found in plants called phytochemicals are fascinating numerous researchers. They are attempting to recognize, measure, and decide their job in human sustenance. This is a perplexing field in its earliest stages. What supplements do wild foods bring to the table?

Since this paper manages verdant greens, all the supplements talked about will just allude to and think about these greens. This thought of contrasting greens with different greens makes its own scope of what thick a high versus a low degree of supplements. The supplement graph gives you a general thought of where supplements are and where they are definitely not. On the off chance that you need to eat nourishments that are supplement rich, you should begin with ones that have a more noteworthy number of qualities which are good for your health. These will be beneficial for you on the grounds that not exclusively will you get bunches of incredible supplements (Tardío, J., Pardo-de-Santayana, M., & Morales, R., 2006).

*The Mediterranean wild food diet:* I have been dazzled by the conventional eating regimen of the individuals on the Greek island of Crete. The Mediterranean Diet, it got known, was an amazingly differing diet because of all the wild foods expended, both plant and creature. The omnivorous eating regimen and incredible way of life of the Cretans came about in long solid lives with basically no coronary illness and little malignant growth and other degenerative illnesses ordinarily connected with present- day living. The vast majority of the wild foods canvassed in this book are eaten as a typical course of each day life during the time they were examined. Those equivalent wild foods are found directly here in our own North American lawns. Practically all the ongoing supplement and phytochemical information in regard to wildwild foods have risen as an immediate aftereffect of scientists examining the wellbeing advantages of the first Mediterranean eating regimen.

*Verdant greens and different vegetables:* In some structure or another, greens were for all intents and purposes some portion of each feast, nearly every dish, for the Cretans, As Diane Kochilas wrote in her book The Glorious Foods of Greece (2001), ''Greens and olive oil were devoured in stunning amounts," Antonia Ttichopoulou, a functioning analyst of the eating regimen, composed that Greeks have been portrayed as having one of the most elevated utilization of vegetables in the western world. These wild foods, a necessary piece of regular day to day existence, were accumulated new ordinarily in season.

Contingent upon their individual qualities, greens were utilized as vegetables, enhancing spices, flavors, and teas. They were eaten new, regularly bubbled, or prepared with olive oil, lemon juice were routinely added to new and cooked greens. There are two ideas to recall;

- Greens were devoured in mind- boggling sums.
- There was an enormous decent variety of species, especially wild species, that were utilized.

Below are a few of the nutrients found in the wild foods. Components that may cause varieties in these qualities are the lavishness or absence of minerals in the soil, the time of leaves at the time of collect, regardless of whether the investigations incorporate just leaves or delicate verdant stems, and whether the examination plants were scavenged or developed in nurseries.

*Omega-3 unsaturated fats:* Fat in the eating regimen is comprised of an assortment of unsaturated fats. Each differently affects the body. These are viewed as basic unsaturated fats because of their functions in development and support with respect to skin, eyes, and the sensory system.

*Vitamin A (RAE):* Nutrient is significant for vision; for keeping up the skin, every mucous film, kick the bucket coating of the stomach related tract; and for appropriate bone development. In wording RAE (Retinol Action Equivalency) a universal proportion of Vitamin A, garlic mustard has the most at

1 ,029 micrograms for every 100 grams of new greens. That is trailed by broccoli leaves, kale, wavy dock, and turnip leaves. A large number of the greens in these outlines have high measures of RAE.

*Beta-carotene:* Beta-carotene has useful properties of its own. It is viewed as a solid cell fortification, may give protection against a few kinds of disease, and may improve insusceptible capacity. Beta-carotene is a forerunner of vitamin A. So, beta-carotene is found in a direct extent to RAE in the diagrams. As with RAE, beta-carotene is found in the most prominent amounts in garlic, mustard, broccoli leaves, kale, curly dock, and turnip leaves.

*Vitamin E:* Vitamin E as alpha- tocopherol is viewed as an amazing cell fortification. The vitamin E assists with balancing out cell membranes; to secure the lungs against ecological harm; to restrain tumor development; to secure the eyes, skin, liver, breasts, and different organs from harm; and to help in keeping up the organic respectability of Vitamin A.

*Vitamin C:* Vitamin C is a significant cancer preventive agent. It boost the production of vitamin E and iron assume a significant function in collagen arrangement in ligaments, scar tissue, bones, and teeth; is significant in amino acid digestion and hormone amalgamation; and assists o with overseeing cholesterol. Here, the best four plants are wild foods, beginning with garlic mustard at 261 milligrams for every 100 grams of greens. These are on the higher scale, since the suggested daily intake ratio is just 73-90 mg for adults.

*Riboflavin:* Riboflavin, otherwise called vitamin B2, is significant for energy creation, unsaturated fat formation, amino acidacidcombination, and development by and large. Wild spinach is the most elevated source of riboflavin at .37 milligrams for every 100 grams of greens. This is trailed by dandelion, developed spinach, kale, and collards.

*Folacin:* Folacin, otherwise called folic corrosive, is fundamental for amino acid digestion, for DNA development, and for the assembling of neuro transmitters. Turnip greens and spinach at 194 micrograms for each 100 grams of greens have the highest folacic concentrations, followed by mature mustard and collard greens.

*Calcium (Ca):* Calcium is generally significant for bone upkeep, nerve transmission, muscle constriction, and blood coagulating. Wild carrot tubers are the richest source by a wide margin at 450mg per 100 grams of new material. That is followed by green amaranth, wild spinach, shepherd s satchel, and garlic mustard.

*Iron (Fe):* Iron is fundamentally utilized in hemoglobin and is engaged with enzymatic movement. Iron is the key for oxygen being provided to the cells, the best seven verdant sources of iron are wild greens. The most elevated source by a wide margin is wood tawny at 8 milligrams for every 100 grams of new greens. The following six are green amaranth, shepherd's satchel, garlic mustard, dandelion, chickweed, and barbed sow thorn. Iron from plants is ineffectively retained.

*Zinc (Zn):* Zinc is a cofactor for more than twenty enzymatic responses, which incorporate alcohol detoxification, bone digestion dietary protein processing, and energy creation. Zinc additionally fills in as a cell fortification. The main seven green sources of zinc are wild foods. Garlic mustard has the most at .91 milligrams for each 100 grams of greens.

In addition to the above-named vitamins and nutrients, there is an addition of manganese, selenium and copper. Wild foods present an array of different nutritional value to the body but so does domesticated foods and vegetables. This all boils down to the handling of the foods and the quality of the foods presented to us.

## 1.6 OXALATES AND NITRATES

Plants ingest nitrates from the soil and process them to frame plant proteins. On the off chance that plants assimilate abundance nitrates and are devoured by domesticated animals before they are changed over to proteins, nitrate harming can happen. Scrounge crops that are over prepared before being collected or brushed can be a typical reason for nitrate harming. Be that as it may, abundance nitrate collection likewise happens promptly in some normal field weeds. Nitrate fixation can shift broadly among plants and developing conditions. Nitrates are most noteworthy in plants in mornings and nights, and on cool, shady days (when plant digestion is slower). Dry spell, preparation and supplement insufficiency can bring about nitrate aggregation in plant tissues. Most noteworthy fixations happen by and large in stems, as opposed to leaves, flowers or organic product/seed. Animal metabolism changes over nitrate (NO3) to nitrite (NO2), which is toxic. This happens less now and again for horses which don't promptly change over nitrate to nitrite. In little amounts, nitrates are decreased by useful microorganisms in the rumen to microbial proteins (Sundriyal, M., & Sundriyal, D. C., 2001).

It is the quick ingestion of enormous amounts of nitrates that can prompt harming, overpowering the rumen's capacity to change over nitrates into proteins. Expanding the sugars (vitality content) in a creature's eating routine can forestall harming as it permits the transformation of nitrates to proteins to happen all the more rapidly, in this way lessening the probability of nitrate harming. Side effects of nitrate harming incorporate sleepiness and shortcoming followed by strong tremors, expanded heart and respiratory rates, imbalance and lack of posture (failure to stand upright without help). Sub-deadly portions can cause premature birth and diminished milk creation. Creatures associated with having nitrate harming ought to be kept calm and the speculate food source evacuated. Scrounges expected of being high in nitrates, particularly on the off chance that they have been intensely prepared with N-compost or experienced dry spell, ought to be tried. Retaining in the abundance of nitrates, a few plants store high amounts of potassium and sodium oxalates (salts). In the event that enormous amounts of oxalate gathering plants are eaten, the rumen is overpowered and unfit to use the salts and they are assimilated into the circulatory system. In the circulatory system, they complex insoluble salts that accelerate in the kidney, causing kidney failure. Sheep are generally vulnerable, at that point cows. Cows can detoxify enormous amounts of oxalates in their rumen, diminishing odds of harming. Creatures can build up a capacity to bear oxalate collecting plants by working up the convergence of oxalate-corrupting microscopic organisms in the rumen. Whenever eaten in limited quantities after some time, with other feed to weaken the focuses in the rumen, oxalate aggregating plants stop to be an issue.

Oxalates otherwise called oxalic acids, are compounds found in pretty much every living thing, in water as sodium oxalate it's commonly found in considerable amounts in marine animals in the form of (Oxaloacetate), and the purslane family which are nourishments known to have high concentrations. Oxalates are an ordinary piece of the human eating regimen and are a typical piece of human amino acid and ascorbic include digestion. You realize a food like spinach is high in oxalates in light of the fact that your teeth get this interesting sand- papery feel when you eat it. As right on time as 1973, the National Research Council of the National Academy of Sciences thought of certain ends (Fassett, 1973) about oxalates in human digestion, as expressed in their report "Toxicants Occurring Wildly in Foods." What follows is my point of view on the oxalate issue dependent on data from that report and my general information on human physiology (Sundriyal, M., & Sundriyal, D. C., 2001).

Oxalates can mix with calcium and different minerals in the stomach tract, making both the calcium and the oxalates inaccessible for use by the body. This may be a defensive adatation that calcium gives to reduce the measure of oxalates consumed. In an ordinary sound eating routine by a typical human, practically all dietary oxalates are discharged with the feaces, with a small part ingested through the intestinal films, it is intriguing to note also that plants with high oxalate content are regularly high in calcium. Along these lines, dietarily, there would be no total deficit of calcium even when consuming oxalate foods; the plant supplies the calcium so that isn't exhausted in your eating routine. In the event that you have an irregular physiological penchant for kidney stones or stomach related problems, see how your body responds when you do eat high oxalate nourishments, such as wild spinach.

Nitrates are a class of substances containing nitrogen and different atoms. In thinking about a yard, a nursery, or a ranch, you should realize that nitrogen is a significant piece of manure. Plants use nitrogen for development. Nitrates are building blocks for DNA, amino acids, and proteins. Without nitrates, development stops in plants and people. People and domesticated animals are diverse physiologically and dietarily. People have diverse eating habits, so any one substance found in a specific food gets weakened. A wide food and diet variety is the best barrier against modest quantities of possibly harmful ingredients in what we consider to be ordinary foods. Diet variety incorporates both the sorts of nourishments you eat and wild foods which are gathered. By and large, there has been no noticeable danger of nitrate harmfulness to grown humans or even youngsters more than 4 months old. Apparently, people four months old and older have nothing to fear and can disregard admonitions about nitrates (Sundriyal, M., & Sundriyal, D. C., 2001).

There have been few examples, just a bunch over the most recent 100 years, of recorded poisonings of newborn children due to ingestion of high nitrate spinach that was made into infant food. The irregularity makes one imagine that the high concentrations of nitrates were an unfortunate and uncommon blend of conditions, that the babies were not given a sufficiently varied diet regimen, or possibly the newborn children previously had poor health. The fact of the matter is that nitrite poisoning is uncommon to such an extent that we ought not stress over it. Being informed is much better than anxiety. The first side effect of nitrite poisoning is methemoglobinemia in newborn children and a flushing of the face and finger/toe points, cerebral pain, and stomach upset. Higher, progressively risky levels can create greater levels of stomach upset, queasiness, heaving, a pale blue shading of the skin and possible breakdown. Once more, remember that while methemoglobinemia is hypothetically conceivable, dangerous nitrite levels have just been recorded a couple of times over the recent hundred years.

## 1.7 AGRIOTROPHYTOLOGY

It is the Study of Wild Food Plants. Many individuals in Africa, South America, and the Mediterranean despite everything gets a considerable extent of their eating routine from wild plants. The now-popular Mediterranean eating regimen included a large portion of the wild greens shrouded in this book. Right up 'til today, the occupants of little Greek towns ear a lot of wild foods. Chronicled agriotrophytalogy includes the investigation of the put down account identified with consumable wild plants. Both logical and lay written works contribute data about wild food uses and food ways. Wholesome agriotrophytology includes supplement examination of wild foods and the functions those supplements, and different synthetic concoctions play in human sustenance. Ethnographic agriotrophytology includes the documentation of conventional wild food courses through observational what's more, talk with strategies. This depicts in detail what is finished by people or gatherings with regards to their circumstances and lives utilizing set up ethnographic techniques. Quantitative agriotrophytology includes the assessment of the utilization and the executives of consumable wild plant assets by people and gatherings. This analyzes the practices and elements required to oversee satisfactory maintainable harvests after some time in various environments, this can be a characteristic and heightened circumstances (Stein, 2008).

Exploratory agriotrophytology includes testing techniques and strategies for social occasion, moving, preparing, safeguarding, and cooking wild foods. Those strategies and techniques can be creative or dependent on precise or best-surmise records of social food ways.

Culinary agriotrophytology includes the science and specialty of transforming wild plants into Food, this incorporates consolidating nourishments, overseeing flavors and surfaces, and creating plans. Rural agriatrophymlogy includes pass on innovation important to domesttticate, develop, collect and transport wild foods to business sectors.

## 1.8 FINDING YOURSELF AND SOCIETY

Making new markets would achieve numerous things. Collecting wild foods would give the accompanying:

- Another device would be added to the arms stockpile of coordinated weed administrations Many weeds would trans-structure from an obligation to an advantage.
- An increasingly differentiated yield level would be available, thus diminishing reliance on only a few plants. In the event that specific crop yields decline, others would thrive.
- The farmer would profit financially . The cost of hand-weeding would be counterbalanced by the pay of collected culinary weeds. Less expenses would be spent on killing or evacuating weeds.
- The proceeds derived from selling the wildwild foods could cover farmhands payrolls .
- Populations will benefit from a variety of assorted foods. This will be useful for both culinary and health purposes.
- Harvesting edible weeds would be useful since there would be no need for herbicides that filter into the groundwater, stay as buildup on plants or affect farmworkers' health.

For this to work harvesters would need to be skilled in the areas of reaping, bundling, and transporting new crops. Wild foods speak to a colossal undiscovered asset that could, under the correct conditions, improve the food alternatives of shoppers, help the farmer, and improve the condition of those in hunger.

# 2

# *Nutrient and Anti-Nutrient Content Of Wild Food Plants*

## CONTENTS

## 2.1 INTRODUCTION

The gift of plants by nature is a resource for human survival. Plants generally occupy an important position when it comes to the livelihood of both humans and animals, some of which are domesticated by people. Plants play the role of being used as food to humans and also other animals that exist on this planet. They are also important when it comes to the regulation of the water cycle, they are also important when it comes to the provision of shelter and raw materials used in different industries. They have also played a key role in traditional and modern medicine. When it comes to food, plants can be defined as edible substances that are useful in the provision of essential nutrients that are important for normal human functioning.

Plants can generally be sourced from either wild or cultivated sources. Wild plants occupy different natural ecosystems, while cultivated or domesticated plants are those plants which were once wild but later bred from their wild habitat and used for subsistence. According to the food and agriculture organization, wild plants can be defined as species of plants that are self-reliant and grow spontaneously as they maintain self-populations within both wild and semi-wild ecosystems and are independent of human action in a manner that is direct (Richard J. Medve & Mary Lee Medve, 2010).

Food gathering resulted in domestication and went on to result in cultivation. This makes it difficult to make distinctions between wild plants and domesticated plants as plants utilized also vary from community to community and also at a global scale. Many plant species that are domesticated are cultivated widely cultivated for example the pistachio which at the same time can be collected from the wild. There are also medicinal and aromatic plants such as thyme, rosemary, oregano, sage and the African stinkwood which have also been domesticated but are sometimes still harvested from the wild. From the differences some anthropologists have come to the conclusion that wild and domesticated is a culturally distinctive term and will have different definitions for different cultures and communities. Despite all these it is still important to understand the value different plants have and how human beings can benefit from their consumption or their utilization (Richard J. Medve & Mary Lee Medve, 2010).

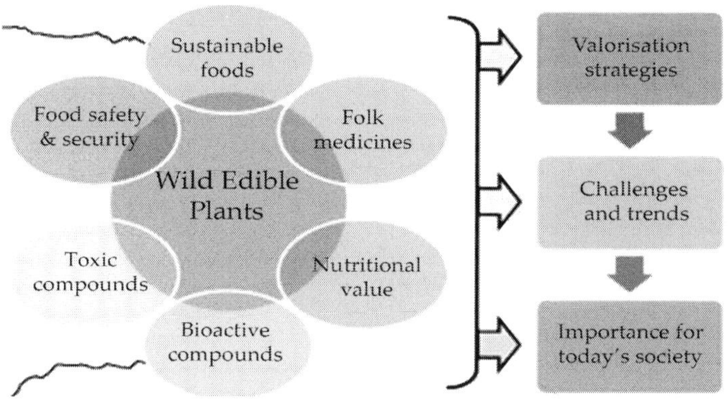


**Figure 2.1.** The benefits of wild edible plants and effects of their consumption.

*Source:* https://www.sciencedirect.com/science/article/pii/S0278691517306191

Ethnobotany as a field of study has been able to develop research in this area and identify wild plant species that are consumed together with other food sources in developing and developed countries. Edible plant parts have been consumed as part of human nutrition since time immemorial and some of the plants are also considered for use in the treatment of illnesses and ,therefore, they benefit humans in several ways. Domesticated plants generally provide important nutrients such as carbohydrates, fats and crude proteins that form basic parts of a diet for many communities.

According to the Food Agriculture Organization, approximately 1 billion individuals globally are assumed to have consumed wild food in their diet at some point. Wild plants can be an important source of vegetable fruits, tubers, cereals, nuts and in balancing nutrition and dealing with food insecurity all over the world especially in countries that are still developing who experience a lot of cases related to food insecurity. Numerous research and written publications provide information on one plant species in various parts of the world and Africa showing the important components of African diets and during drought periods (Saunders, 2012).

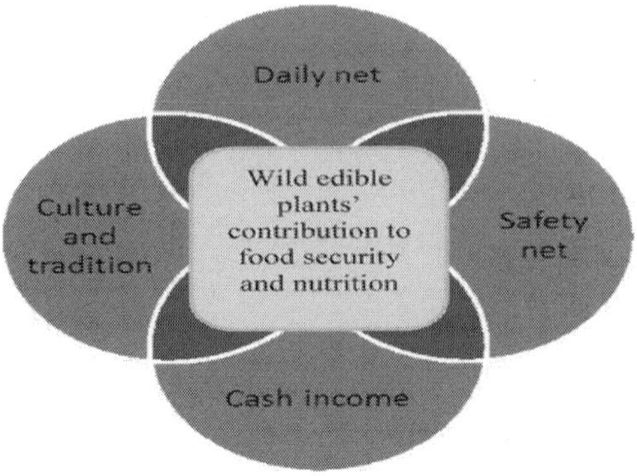


**Figure 2.2.** The contribution of wild edible plant foods to food security and nutrition.

*Source:*https://agricultureandfoodsecurity.biomedcentral.com/articles/10.1186/s40066-017-0097-5

Nutritionists and dietitians have recommended diversification as a way or improving nutrition status and also dealing with food insecurity. Agricultural food systems continue to put more emphasis on diversification of foods rather than dwelling on certain food crops which are cereals. Plants are major sources of nutrients and their consumption of wild plants is becoming more and more acceptable with time and have the potential to solve some of the issues with regards to food insecurity in many countries especially from developing countries. Plants including wild plants are major sources of many active substances ( phytochemicals), which are reported to be a key to good health and it commonly differs based on the plant family.

Researchers have investigated many wild plants for different nutrient and anti-nutrient components. However, their composition of the mentioned phytochemicals differs depending on the plant family and is affected by factors such as the characteristics of the genotype, climate, edaphic factors and how well these plants are managed. For example, research has shown that such elements are plant-specific and the mineral nutrient content together with the phenolic compounds and antioxidants will differ depending on how different nutrient sources are applied to the plants or in the case of wild plants, which nutrients they have access to from the soil (Saunders, 2012).

It is important to note that as much as the different classes of chemical compounds that add significance to the health of human beings, the others that can be very toxic and also cause harm to human beings upon consumption. This means that phytochemicals can be classified according to nutritional and anti-nutritional factors. Nutritional factors can be classified into essential fatty acids protein vitamins minerals and phenolic compounds well anti-nutritional factors can be classified into oxalates, tannins, nitrates, phytates and so many others.

Apart from the provision of these biomolecules, plant species play an important role when it comes to their achievement of good nutritional status. Micro and macronutrients together with heavy metals can have different risks on the human body. Plants have been acknowledged as an essential cheap and creative source of proximate nutrients, minerals, vitamins ,as well as antioxidants, for poor communities. However, the same dangers of these anti-nutritional factors are reported by scientists carrying out different studies in the case of many traditional or wild plants. The anti- nutritive components that exist in wild plants and plants, in general, have the potential to influence the bioavailability of other nutrients especially those that are important for the proper functioning of the human body. Due to their bioactive and physiological properties, some chemical components of food have also been acknowledged as having a protective effect and when combined with plant nutrients they can offer protection to the body. Chemical components in plants can also form complexes with nutrients that exist in plant foods and can render some nutrients that exist in plants as unavailable. The ability to bring out these effects can be brought out depending on the concentrations that exist in the food plants (Press, 2013).

It is important for researchers and similarly those who consume wild plants to understand the importance of these chemicals and their effects on human health and the methods that exist when it comes to assessing they are content in wild plants. The case of the existence of these biomolecules in wild plants should be made aware to people who utilize such plants as sources of food.

## 2.2 WILD PLANTS DIVERSITY

Globally, many plant species have been used by human beings and have been considered as valuable resources. Almost 5000 species have been cultivated of the 250000 species that have been identified and only a few of these species have been domesticated and have been traded globally. Only about 103 plant foods contribute to 90% of the national per capita supplies, about twenty to thirty species have also been referred to as staples that supply nutritious foods to the human population. On the other hand, many plant species have been grown locally, are scarce or have undergone partial domestication and many are still being gathered from the wild for human use (Press, 2013).

**Figure 2.3.** Edible plants from the Brassica family both wild and cultivated.

*Source:* https://www.researchgate.net/figure/2-Plant-morphological-diversity-of-Brassica-wild-species_fig2_225208492

The partially domesticated or plant foods collected from the wild are usually found within the tropics for instance, the Plant Resources of South East Asia have projected about six thousand plant species and its basic list of species used by humans in that area and assumes similar levels in other areas in the tropics. A figure of eighteen thousand to twenty-five thousand plant species has been estimated to be used for medicinal purposes globally. This does not consist of most of the twenty-five thousand plant species that are used as medicine in many parts of the world such as China, tropical Asia, South America, India and Africa.

Wild species used by human beings occur in biogeographical regions and different habitats.in the different areas, human beings have learned to make use of the plant diversity that exists no matter how harsh the environs are, for instance, a large population depends on the diverse vegetation from forests, woodlands, thickets, grasslands and swamps. Even in the montane regions such as the Andes, specialized systems have developed in which people are able to exploit and make use of wild plant species (Pardo-De-Santayana, M., Tardío, J., & Morales, R., 2005).

In desert and arid areas, wild plant species play an important role especially with regards to pastoral and agro-pastoral communities that are dependent upon the availability of vegetation in wild ecosystems for grazing or browsing when it comes to animals that are domesticated. Tropical forests have also been acknowledged as home to a huge percentage of the plant diversity on the earth and these plant species are a major source of food and other food- related products. The communities that live around the Amazon has more than fifty percent of its rural population depending on forest- related activities, including their construction and the survival of human beings. Worldwide, millions of indigenous communities live in tropical forests and the use of wild species from them form an important part of the household economy but the information differs within the regions and among them.

## 2.3 EDIBLE WILD PLANT SPECIES

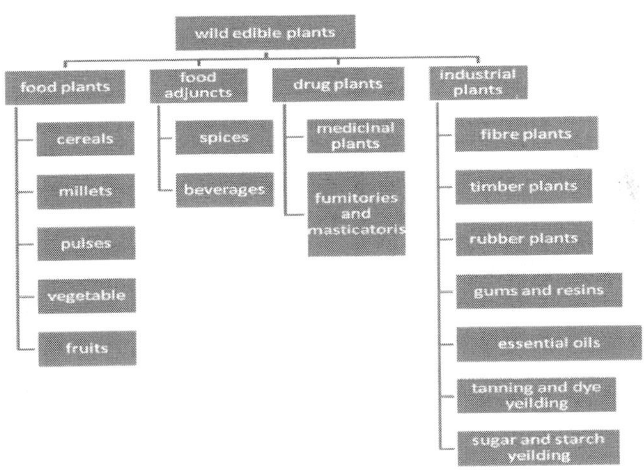


**Figure 2.4.** Different categories and uses of wild plants.

*Source:* https://link.springer.com/chapter/10.1007/978-3-319-63037-3_3

Consuming wild plants by humans is not a new concept and has existed in many communities but it continues to be explored. Despite this, there isn't a lot of information on edible wild plants but literature can be traced from their past about their use using skills from the field of ethnobotany and knowledge from archaeology and anthropology. Although most plants in the contemporary world have been improved in terms of their quality for through bio-fortification, there are individuals and communities who still have a preference for plants from the wild.

Edible wild plants can also come in the form of weeds or bushland forest species and have received little attention from scientists especially as they evaluate and decide on issues associated with development in agriculture and nutrition. Research has also shown that the dietary value of wild plants is not far from plants that are cultivated (Pardo-De-Santayana, M., Tardío, J., & Morales, R., 2005).

### 2.3.1 Wild Edible Plants Consumed at Present by Native Societies

It is assumed that wild plants have been consumed in significant amounts in the past compared to the amounts that are currently consumed. They are a wide range of plants that are consumed from the wild for example those from *Portulaca, Chenopodium* and *Sonchus and* wild plants from *Amaranthaceae* family. The numerous number of wild plants are suitable for use as vegetables and in rural areas it can be pegged as one of the reasons why individuals are not interested in cultivating their own vegetables. This is because vegetables are often purchased in the market while wild plants will not require any effort in its cultivation and also not monetary resources can be put into cultivation or purchasing. This is a sufficient reason for the utilization of wild plants and also that they will still fulfill some of the nutrition requirements. It is also easy to plant for crops for their leaves rather than any parts of the plant such as fruits, roots and other parts. The leaves are also rich in a number of nutrients compared to other plant parts.

**Figure 2.5.** Amaranthus retroflexus

*Source:* https://en.wikipedia.org/wiki/Amaranthaceae

Edible wild plants with high nutritive value have also been utilized in Asian countries like Japan, India, China and Korea and other parts of the world like South America and Africa and it becomes complicated when it is time to evaluate all the plant species that are used globally and ,therefore, estimations are made to this regard. Currently, the approximate value is about 150 out of an estimated 30,000 edible plant species that provide the world with food.(Shelef, O., Weisberg, P. J., & Provenza, F. D., 2017).There are many plant species that have the potential for human consumption if well explored and studied by assessing their nutritive value, whether they are poisonous or have anti nutritive components that can suppress other nutrients.

*Asia:* In Asia, many plant species are used as vegetables, some of which are regarded as novel foods or as exotic plants in countries that are more industrialized or developed and sometimes either grow in the wild, fields as weeds. Some of the vegetables reported to be used by individuals and communities in South East Asia include *Coccinia indica, Ipomoea aquatic, Ipomoea batatas, Morinda citrifolia, Moringa oleifera, Neptunia oleracea, Centella asiatica, Momordica charantia, Basella alaba, Basella rabra* and *Gnetum gnemon.* Their nutrition values have also been assessed and recorded.

China is one of the largest and oldest countries and can be considered a centre for vegetables globally. Some statistics have shown that it is home to almost two hundred and thirteen families, eight hundred and fifteen and one thousand eight hundred and twenty two species of plants that are used as vegetables in China. Approximately five percent of these plant species have been commercialized and are traded at different levels. The nutritional value of twenty five unusual Chinese vegetables has been provided by researchers. Some of the plant species are rich in the nutrition components for example, *Mormodica charantia*, *Ipomoea batatas*, the *Amaranthus* species and many other plant species.

The Xioxing'ang Mountain is found in the eastern mountain region and is one of the major timber bases in the country. The deposits of wild fruits in this area are usually huge. A number of wild fruits are also distributed in this area and are also easy to collect. Lihuan and Ling in 1999 carried out research on wild plants in this area and mainly focus on thirteen genera the *Rosa, Actinidia, Ribes, Rubus, Colyrus, Pyrus*, and others. Most of these wild plant species had significant amounts of vitamins for example, plants from the *Rosa* species had exceptional amounts of ascorbic acid.

Leaves of plants are also in food preparation in India and research was done for these plants in order to assess their nutritional value in relation to essential fatty acids and the presence of tocopherols. The report was done on plants like *Tamarindus indica*, *Oxalis corniculata*, *Rumex vesicarlus, Althernathera sessilis, Ttigonella feounumg* and the *Talinum triangulare.* These plant species can either be cultivated or extracted from the wild but most of the time people get them from the wild (Angier, 2001).

*Chenopodium* wild plant species has been widely used in India. From more than two hundred plant species, only two to three species have been successfully cultivated. In India, *Chenopodium album* has been widely used but is common in the Himalayan region in India and can also be found in the plains of Punjab and similar to the South. It is a common plant species among the people of the Western Himalayas and several areas. The wild plant is utilized as food, it is also used as a herb and as a grain crop. It is a pseudocereal and it appears to be a good supplemental protein to the diets of those who consume them as cereals.

*Amaranthus* is also an important plant genus in Asian countries and is commonly consumed by Indians. It is considered a good source of protein and it also affordable and is an important source of Vitamin A. it has high nutritive value, it is also tolerant of arid environmental conditions and it can also grow in areas with poor soils. In India, the Garhwal Himalaya, there is an abundant growth of wild fruit trees particularly during the summer. The local consumption of these wild fruits was looked into that is *Aegle marmelos, Berberis asiatica, Hippophacea rhamniodes, Myrica nai, Rubus ellipticus* and *Prunus armeniaca.* Various fruits are also consumed raw by local communities in these areas and are rich sources of proteins, carbohydrates and a range of micronutrients and can be regarded as alternate sources of food.

In the tropics, the underground tuber *Discorea* species is one of the most important carbohydrate sources. Many wild tubers have shown high nutritive value and also have a high protein content. Other wild species that can is consumed by many individuals and communities include the *Discorea alta* also has high protein content and can be regarded as highly nutritious food. Other important plant species in India include the Bamboo species. The seeds of some plants from this species are consumed often especially by the poor during periods of drought. They can also be used in the production of beer. Other species are also used to feed livestock for communities with these plants in their environment. When the species were analyzed and their nutritive value for both micronutrients and macronutrients. They were found to have significant amounts of ascorbic acid, carbohydrates, proteins and minerals (Angier, 2001).

*North and South America:* In North America, the native Indian tribe of Nuxalk makes a good example when it comes to studying ethnobotany related to the use of wild edible plants, this community made use of several wild plant species. During spring, green plants are first harvested and similarly are used fresh during the season. When in plenty, they are stored to be used later. They have made use of green plants and also trees as both sources of food.

In Brazil, some widely used plants have also been studied for their composition in relation to their vitamin A content that is *Amaranthus viridis, Lepidium pseudodidymum, Xanthosoma* species, *Sonchus oleraceus* and *Portulaca oleracea*. From the study, they came to the conclusion that native wild leafy wild vegetables are rich sources of provitamin A compared to other grown vegetables that they investigated. They also justified the need for their cultivation commercially as they were not difficult to grow, their availability was throughout the year and so they were also acceptable to the communities in which they existed and could be and it also has the potential to reduce Vitamin A deficiency.

As the world population continues to grow, and food continues to become scarce, the consumption of plant proteins has often been proposed. Proteins sourced from leaves can be cultivated in tropical regions and could have the advantage of exploitation all year round, with consistent amounts of sunlight. The protein and amino acid content from the leaves of twenty-three wild tropical plant species have been investigated in the United States and the plants include *Cajanus cajan*, *Momordica charantia*, *Portulaca oleracea,Ricinus comunis*, *Hibiscus sinensis, Tamarindus indica* and other plants. The researchers concluded that the above -mentioned plants had significant amounts of essential amino acids.

Plant samples of edible plants of Guyana which are less known went through chemical analysis. The results of this analysis indicated that some of the plants have the potential to be good sources of edible oils and they also represent a group of species that can provide adequate nutrients without using a lot of resources (Ballerini, 2012).

*Africa:* Countries like Mozambique in Africa have had the wild plants used by native communities surveyed. Edible leaves used throughout the country have also been surveyed and some of the plants include *Amaranthus caudatus*, *A. graecizans*, *A. spinosus*, *Corchorus trilocularis*, *C.tridens*, *Momordica balsamina*, *Colocasia antiquorum* and *Bidens pilosa*. The fruits from *M. balsamina* are used together with leaves and also combined with nuts to form a traditional meal commonly referred to as Canaca. The researchers who did the study came to the conclusion that it would be a good supplement and a good way of improving cereal- based diets such as maize without completely change food habitats since the plants that were investigated were located in different areas within the country.

Wild and cultivated species of *Amaranthus* are eaten all-over Africa and are sometimes grown as grain crops in Latin America and in India. Many plants from these species are known for producing grains rich in protein and are therefore utilized as plants that are cereal-like. The use of these seeds is almost the same as the use of wheat as it is grounded to form flour which is then made into dough. These plants show promise when it comes to the improvement of the quality of life in tropical areas.

The leaves of *Pterocarpus mildbraedii* are used in preparing traditional soups in Eastern Nigeria. This tree vegetable does well in forest environments. The leaves have high nutrient content compared to steamed spinach. Other edible trees have also been proposed for food and this is as are we to try and avoid the cutting down of trees.

Yams are traditional food crop and are originally in Africa America and Asia but currently, it is distributed all throughout the tropics and a few of the species can also be found in temperate regions. Research has shown that's some local wild yam species have been found to be very nutritious but still have some toxic principles. Despite containing anti-nutrients such as alkaloids yams have the potential to serve a security source of nutrients (Ballerini, 2012).

The genus *Nymphaea* is one of the largest genera of water lily family and has more than 50 of its species drawing in tropical and temperate climates. For which species are found in Malawi that is his species *N.caerulea, N.capensis and N.lotus* and *N. pertasiana* which is gathered from the Southeastern part of the country when was analyzed for its nutrient content and anti-nutrient components. The study was found to have several important nutrients for example iron and protein.

Analysis of wild plants has also been done in southern Sudan and the plants were analyzed for the presence of beta carotene. Of the plants analyzed plants, 14 were found not to have significant contents of beta-carotene go compare two plants that were being cultivated. Analysis of indigenous vegetables has also been done in Kenya at Machakos district the analysis was done to determine carotenoid levels. From the said it was found that most of the plants were able to contribute more than 50% of the daily dietary requirements for vitamin A and C.

In Niger, native plants have continuously become a significant staple in the event where harvest from cereals is not adequate enough to support populations that live in some regions in Africa. Research has shown that famine foods derived from the wild and cultivated in Niger why rich sources of protein and amino acids, lipids fatty acids beta-carotene vitamin A and trace minerals some of which included the *Moringa oleifera*, *Leptadenia hastate*, *Amaranthus viridis*, *Idonsia digitata*, *Ximenia americana* and others. From the study, conclusions were made that the plant species had significant amounts of minerals such as copper and zinc, they also had significant amounts of essential amino acids and fatty acids. The fruits and leaves of 7 native plants that are *Idonsia digitata*, *Balantines aegyptiaca*, *Boscia senagalensis*, *Cassia obtusifolia*, *Ceratotheca sesamoides*, *Leptadenia hastata* and *Ziziphus mauritanica* commonly consumed in northern Senegal were analyzed for their carotenoid levels. World's plant species have also been established to offer various nutrients for people living in sub-Saharan Africa.

A cultural, ecological and Nutritional study was done in Swaziland on wild edible plants. Ever data from this study established that the wild edible plants are good sources of proteins and minerals and were mostly eaten by indigenous groups in the country. The plant investigated in the study were mainly leaves and fruits. The species investigated included the amaranth species, the *Corchorus* species, *Chenopodium* species, the *Hibiscus,* the *Mormodica, Portulaca* and *Sonchus.* In countries like Egypt, fresh vegetables are available throughout the year. In the country, researchers have studied and analyzed wild plants to determine their vitamin composition ,for example, in some species like the *Colcosia esculenta*, *Corchorus olitorius*, *Malva persiflora*, the Rumex species, *Sonchus oleraceus* and *Trigonella foenum-graecum.*

*Australia and pacific:* Brand at al. did an analysis of the nutrition contents of some wild edible plants in Australia four plants that produce seeds and nuts. Seeds and nuts were an important part of the diet of the Aboriginal community in Australia before modern civilization. Which community gathered seeds as one of the major activities Darling river basin. An analysis was done for 21 seeds from different plant species including the acacia species that is *A. coriceae*, *A.murayama*, *A. tenuissima*, *A.terminalis* and *A.victoriae*; two plants from the species Portulaca that is the *P.oleraceae* and *P. intraterranea*; two plants from the Grevillea species that is the *G.leucopteris*, *G.pteridifolia*; *Nelumbo nucifera*, *Pandanus spiralis* and other plants. They made the conclusion that most of the seeds why rich in nutrients especially proteins fats and minerals (Stein, 2008).

Most of the Pacific foods have not been analyed in order to determine the nutritional composition and the Aprogram Pacific Island Food Composition program has tried to deal with this inadequacy by carrying out an assessment of some of the plants. Plant samples of *Hibiscus manihot* who included and we extracted from three different areas. The plant was able to show almost all nutrients. It was able to demonstrate its superiority to temperate vegetables such as the Chinese cabbage which is mostly consumed by individuals living in the Pacific islands. The researchers came to a conclusion after assessing different samples and to them the observed difficulty in obtaining a single reliable nutritive value for inclusion in a football composition table.

*Europe:* The first farmers in Europe mostly consumed roots, stems, leaves and seeds of different plant families and in the contemporary world most of them are no longer considered for nutrition. The seeds belonging to some plants which were considered edible now have been extracted from archaeological sites. Some scientists have studied vitamin C and A contents in some wild edible plants originating from Italy. Some of these include plants that are grass like for example the *Amaranthus retroflexus*, *Chenopodum album*, *Diplotaxix tenuifolia*, *Plantago major*, *Portulaca oleracea*, *Sonchus tenerrimus* and *Taraxacum officinale*. The plants were discovered to have high vitamin content.

Edible wild plants have also been studied in Germany especially in the determination of vitamin c content that is by Franke and Kensbock. They carried out analysis of 55 edible wild plants and looked for the USA's salads. Some of the plants were vitamin C. They came to the conclusion that vitamin c who's higher in edible wild plants compared to those plants that were being cultivated. Plant leaves roots and fruits from the plant *Crithmum maritmum* ,which is commonly consumed by local communities in Europe especially in Italy and Spain. It is consumed raw especially in salads but it can also be pickled and it has significant amounts of vitamin C (Stein, 2008).

Nutrition content of some leaves of the *Chenopodium* species gathered from various areas of southern Spain have also been analyzed. Results from this study who are able to conclude that the plants were important when it comes to providing good nutrition since they had significant amounts of proteins, minerals and vitamin C in the three species but who investigated that is the *Sonchus* species, *Crucifer*a species. Other wild plant species were also included in the study as alternatives to green leafy vegetables that are being consumed for example the *Amaranthus*. The toxic anti-nutritional factors also included when analyzing the different plant species but were in low amounts (Stein, 2008).

## 2.4. NUTRITIVE AND ANTI-NUTRITIVE CONTENTS OF WILD EDIBLE PLANTS.

### 2.4.1 Nutritive Factors of Edible Wild Leaves

*Proximate compositio*n: The nutritive value of various wild edible plants which is used as substitutes for green leafy vegetables can be assessed. For many of these species, the moisture content from the edible leaves ranges from 70 percent to 90%. This varies depending on the stage of maturity of the plant. For example, the *Amaranthus* and *Chenopodium* species seem to have low moisture content whereas species like *Roseaceae* and *Portulacaceae* have higher moisture content.

There's also a variation in the crude protein of the leaf samples that have been investigated for example species from the *Amaranthus* have high protein content and also from *Chenopodum* and *Cruciferae.* Low protein content was observed in plants from there *Rosaceae* and *Umbelifirae*. Carbohydrate content also was sent to vary depending on the plant family. Two plants from the *Corchorus* species had high values, greater than ten percent of the fresh weight. Fiber and that can be determined using various procedures in the analysis. The most frequently used method is the detergent fiber for example the amount of fiber Irish affected by the procedures used to gather the leaves like petiole inclusion from the plants comes from High amounts of fiber. Analysis of fiber content it's also be affected by the physiological stage of the plant. For example, the *Chenopodium* species expressed high fiber content but after drying the differences among the species spread to disappear.

Lipid content on fresh leaves fell between 0.2 and 1.9 grams for every 100 grams of fresh leaves. The lipids found in leaves are made up more of phospholipids and glycolipids then seed lipids which are often made up of neutral lipids. The Ash content for the wild plants that were assessed was able to reach a high value of 5.9 grams for every 100 g in *Parietaria diffusa*. The total energy was also calculated from the values analyzed in terms of proteins, lipids and carbohydrates. The species analyzed had lipid values of 100 to 300 kilojoules for every 100 grams.

*Mineral composition of edible wild leaves:* The composition of minerals in edible wild leaves usually varies depending on the soil type this makes taxonomic classification as a variable that is not important in all cases when it comes to the analysis of minerals. Edible plant species for example *Beta maritama, Crinthmummaritimum* expressed high levels of sodium and this is attributed to their habitat and their halophile nature. Potassium range also varies from plant to plant for example,e the plant *Chenopodium album* had the highest content of potassium that is 1323 milligrams for every 100 grams, species like the *Rubus spectabilis* had 8 milligrams per 100g and in *Parietaria diffusa* had 883 milligrams per 100 grams for potassium content. Magnesium content for all investigated plant species was almost similar to the *Amaranthus* species containing the highest quantities although as observed the *Malva sylvestris* was composed of the most significant amount of magnesium what is 716 milligrams per 100 phosphorus amounts were also consistent among most plant species. Most of the plants had about 16 to 118 milligrams per 100 grams. But for most species, the values laid were between 20 to 18 milligrams per 100 grams. For wild edible leaves, the select trace minerals like iron, copper, and manganese also differed. Species like *Cruciferae* and *Chenopodium* were are the highest when it came to iron content. Copper levels were high in *Beta maritima* and *Urtica dioica*. Manganese was almost the same in most species and ranged between 0.2-1.6 milligrams. Zinc content was observed to be highest in species like *Pterocarpus mildbraedii* but from most species, the range fell between 0.2 to 2.0mg.

*Vitamins of edible wild leaves:* Wild plant foods from the species Amaranthaceae, Basellaceae , Chenopodiaceae ,Compositae, Convolvulaceae, Cruciferae, Malvaceae, Plantaginacea, Poligonaceae, Portulacaceae, Rosaceae and Umbelliferae were select analyzed for vitamins composition of total carotenoids, beta carotene and Vitamin C. the total carotenoids fell between 4.3 to 19 milligrams per 100 grams of plant material. Beta-carotene consisted of less than twenty percent of the total carotenoids in most edible food plant species. Beta carotene was observed to be high in *Amaranthus viridis* and *Sonchus oleraceus*. Ascorbic acid in wild edible plant foods has been observed to be higher compared to cultivated plant species.

Amaranthaceae, Cruciferae and Chenopodiaceae were reported to have exceptional amounts of vitamin C.

*Amino acid composition of edible wild leaves:* The amino acid composition of select wild edible leaves can be done in comparison to the world health organization protein standard which uses the egg white protein as a reference protein. Apart from a few plant species most of the wild edible plants process most of the essential amino acids. Species like Moringa have 100% of the standard in all essential amino acid groups apart from blazing in which the score was 97%. Plant proteins can be of importance in some low protein diets for example in the case where the limiting amino acid is lysine which is associated with cereal-based diet, leaves that contain these amino acids can be supplemental in this case.

*Fatty acid composition of edible wild leaves:* Fatty acid content saponifiable oils can be analyzed four different edible wild leaves. Select species were analyzed and from this, it was established to be around 1 to 2% of the dry weight except for the *Sonchus* species in which the ranges fell between 45% of the dry weight. The oil found in leaves is generally low for most plants. Always in leafy green vegetables are almost similar in exception of *Zizhiphus mauritiana* which expressed significant amounts of oleic acid.

*Nutritive value of edible wild seeds:* When assessing the nutrient content in edible wild seeds, the variation in data is as a result of the previous factors to which they are exposed to. The protein content is high in *Fabeceae* while *Bambuseae* expressed large content when it came to having the largest amount of lipid and seeds from *Portulaca oleracea* produced the largest amount of ashes.

Analysis of mineral content shows that phosphorus is present at all levels and surpasses the content which is usually found in most grains. Calcium and magnesium also found to be high compared to the amounts found in grains that have been cultivated. The composition of fatty acids of some wild species was concerned in some of the wild species like the *Chenopodium album* expressed higher percentages of alpha- linolenic acid and erucic acid. *Gramineae* showed lower values when it came to saponifiable oils.

When it came to amino acid content the *Poligonaceae* and the *Gramineae* species had the smallest amount of total proteins while those from the *Crucifera* family head values higher than fifteen percent. It was also recognized that the amino acid profile s in leaves and seeds was almost similar. Most of the time leaves usually lack some amino acids such as lysine, methionine and tryptophan. Seeds from the *Rumex crispus* had significant amounts of tryptophan, seeds of the *Gardenia lutea* were good sources of lysine while seeds from *Fumaria officinalis* and *Setariaviridis* were reported to have significant amounts of methionine.

*Nutritive value of edible wild fruits:* Wild fruits are rich in their nutrition composition compared to fruits that are cultivated. These wild plants are different when it comes to their moisture content and other nutrients. Selected wild fruits such as *Boscia senegalensis* have significant amounts of protein while other species like *Sapotaceae* were found to have low amounts of protein. *Hippophae rhamnoides* has a significant amount of fat about 25-30 times more than other wild fruits or those that are cultivated. *Tamarindus indica* was found to have a high carbohydrate content. The vitamin C content was high in *Ziziphus mauritiana* and *Adansonia digitata*, while *Momordica charantia* expressed high amounts of Retinol Equivalent. *Adansonia digitata* and *Momordic abalsamina* fruits were exceptional sources of phosphorus and calcium, whereas *Randiagenipoe flora* and *Tamarindus indica* were found to be high in iron.

## 2.4.2 Anti-nutritive Factors

Chemical compounds that are not desirable in plants are referred to as anti-nutrients and they exist in both cultivated and wild plants. These substances can also be called allelochemicals. Their presence in plants in terms of amounts and how they are distributed differs from one plant species to another. Anti-nutritional factors can also be viewed as not being inborn characteristic but rather it is seen as dependent on the digestion process depending on the animal that consumes the plant. The adverse effects of the anti-nutrients will depend on the dietary patterns and the processing means prior to the consumption of the plant food. Some of the anti-nutrients commonly found in wild plants include oxalatyes, nityrates, tannins, phytates and cyanogentic glycosides and they are commonly found in the leaves of plants.

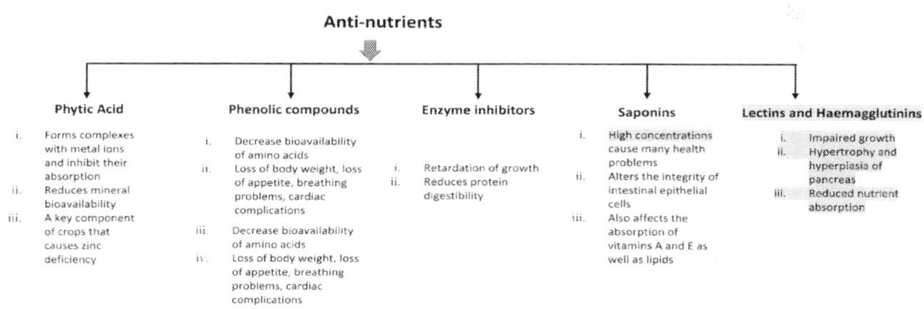


**Figure 2.6.** Different classes of anti-nutrients.

*Source:* https://fppn.biomedcentral.com/articles/10.1186/s43014-020-0020-5

*Nitrates:* Nitrates are some of the most significant wild substances that are commonly found in wild and cultivated vegetables and it is responsible for the characterization of the plant quality. The amounts of nitrates are high in the leaves compared to other plant parts like roots or fruits. They are several factors that affect the nitrate content of plants for example cultivated plants will be affected by the amount of nitrogen fertilizers added and the timing of fertilizers, all the plants will be influenced by environmental and genetic factors. Exploiting the nutrition content from edible wild plants can be inhibited by anti-nutrients such as nitrates and oxalates. Most of the time, wild edible vegetables show high concentrations of nitrate compound.

Amaranthaceae family has plants rich in nitrates. Some parts of edible plants, such as amaranth leaves, have been shown to be a potential vegetable source of nitrate. Vegetables like spinach and beet have been found to have significant levels of nitrates which can cause harm to human beings , since nitrates may act as potential carcinogens after being transformed during metabolism to nitrosamines, changing from nitrites first to nitrite and then to nitrosamines. Salts accumulate in leaves as a result of the addition of excess fertilizers. The *Cichorium intybus* had a concentration of 75 milligrams per kilogram, *Diplotaxis tenuifolia* had an average of 3874 milligrams per kilogram and the *Papaver rhoeas* had greater than two thousand five hundred thousand milligrams per kilogram. *Portulaca aleracea* had values of nitrate ranging between 43.9 mg and 49mg. *Sinapis arvensis* had about 3028 mg per kg and the *Urtica dioca* had nitrate values ranging from 850 to 161mg per kg. (Montefusco et al., 2015)

*Oxalates:* Oxalates are formed when oxalic acids combine with salts or minerals. Oxalic acid is found in the cells sap of most leaves in wild vegetables and this depends on the plant species. Oxalates can be found in different forms for example in insoluble salt forms such as calcium, magnesium and iron. Soluble salts include those formed from potassium and sodium or can form from these two combinings. Oxalic acid does not affect the zinc will be absorbed and metabolized. Oxalates that are not soluble are excreted from the body while those that are soluble affect the human body through the process of chelation with calcium found in the diet or with other minerals making these nutrients not available for absorption and assimilation.

Calcium oxalate crystals are stored in the kidney and can lead to kidney stones which is a serious health problem. How calcium is absorbed varies depends on the calcium and oxalate ratio. Other than through ingesting oxalates through the diet, oxalate formation occurs in the human body as by-product of vitamin C and glyoxylate metabolism. How oxalate is distributed is not even in plants and differs depending on the plant species.

Oxalic acids in wild edible plants are not in amounts that can be toxic. Oxalic acids of more than five grams can be toxic to human beings especially in the case of adults. For instance, when greater than a hundred grams of rhubarb are consumed they are representative of a tenth of the minimal lethal dose. As mentioned earlier the amount of oxalic acid will affect the availability of calcium. Available data shows that the likelihood of low calcium bioavailability was viewed in Amaranthaceae, Chenopodiaceae, Poligonaceae and Portulacaceae. There are also observed similarities in cultivated plant species. *Allium ameloprasium* has oxalic acid `of 6.32 plus or minus 0.65 mg/100 grams and 11.13 plus or minus 0.48 milligrams per

100 grams when two different populations were investigated. For *Cichorium intybus* when two different populations were investigated, the values ranged between 5.67mg to 10.65 mg for a hundred grams. *Foeniculum vulgare* had two populations investigated and had oxalic acid values ranging from 2.29 to 8.63 milligrams for a hundred grams. *Papaver rhoeas* had oxalic acid values ranging from 365 to 462.5 mg per 100 grams. The *Portulaca oleracea* had 1.27 milligrams to 0.55 milligrams of oxalic acid. Other wild plants with oxalates include *Silene vulgaris*, Rumex acetosa and other wild edible plant foods. (Guil et al., 1996)

*Tannins:* The word tannin was coined by Seguin and it meant a substance that existed in plants especially vegetables and played an important role when it came to the conversion of animal skin to leather products. Vegetable tannins are polyphenols with a high molecular weight and form complexes with nutrients such as carbohydrates and proteins in aqueous solutions. Tannins found in wild plants are present in the barks, wood, fruits, pods, leaves and plant galls. Tannins are products of metabolism and are not intrinsic products that result from chemicals being transformed (Stein, 2008).

The classification of tannins can be of two different forms, the first group is classified on the basis of structure hydrolyzed tannins and condensed tannins. The condensed tannins are found in higher groups of plants why the hydrolyzed tannins are made up of glucose circled around by phenolic acids and undergo hydrolysis in their interaction with acids, bases and specific enzymes. Tannins that are condensed are mostly oligomers and are resistant to being hydrolyzed. Tannins in most vegetables are soluble in water and exist widely. Tannins result in effects that are not desirable in food when present in significant amounts. Tannins are capable of forming complexes with proteins via hydrogen bonds and covalent links. This leads to the proteins precipitating hence they became unavailable when it is time to get absorbed. *Papever rhoeas* come from the family Papaveraceae and the leaves and flowers produce tannins, citric acid, malic, tannins and flavonoids.

*Phytates:* Phytic acid is a compound that exists naturally in plants and serves as an important storage for the mineral phosphorus in many wild and cultivated vegetables. It exists in plant tissues in the form of salt cations like potassium, magnesium and calcium. A relevant amount of phosphate groups with negative charges in phytic acid for forming chelates with essential mineral nutrients in the human body reducing their bioavailability. Phytates negatively affect the activity of enzymes in the digestive system and chelate with mineral co-factors or as they interact with protein for example phytates affect the regulation of zinc.

Proteins bind with phytates directly or indirectly via a cation bridge. Techniques used in plant preparation for example by being boiled and cooked do not have any effect when it comes to the reduction of the phytic acid since it is stable in heat. Research has shown that despite different food preparation methods being used to reduce their levels, the most efficient means include lactic acid fermentation and soaking in acidic mediums. The main techniques of determining phytic acid in plant samples were developed by Wade and Morgan and other methods were developed afterward. Modern methods have also been developed such as complexometric titration, through enzyme reactions, potentiometric and ionic chromatography. In a study, the phytates composition of 8 plants was looked into and it was concluded that *Portula caoleracea* had 823mg per 100 grams as the highest value, with *Solanum indicum* had a value of 659mg per 100 g and the *Eulophia* had a value of 255mg per 100g (Angier, 2001).

*Cyanogenic glycosides:* Cyanogenic glycosides come from amino acids and a group of secondary metabolites that exist in more than two thousand five hundred plant species. Cyanogenic glycosides are defined as the glycosides of α-hydroxinitriles. Amygdalin is one of the first glycosides to be isolated. Cyanogenic glycosides are formed from seeds being hydrolyzed for example the seeds of bitter almond *Amygdalus communis*. Most of the time plants have been known for their production of small amounts of cyanide together with ethylene being produced. Cyanogenic glycosides have been investigated due to the poisonous nature of hydrogen cyanide. The family Bambusoideae has about one hundred and ten edible species and contains some amounts of cyanogenic glycosides and therefore boiling should be done before they are consumed.

*Alkaloids:* Alkaloids are one of the largest groups of chemicals produced by plants and most of the time they exist as salts for plant acid such as oxalic, malic, tartaric and citric acid. They have small molecules and exist in about 20% of all vascular plants and have several carbon rings with side chains. Nitrogen molecules are sometimes in place of some of the carbon atoms in the structure. The synthesis of alkaloids occurs by the actions of amino acids. Amino acids that are decarboxylated lead to the production of amines and upon their interaction with amine oxides aldehydes are formed. The heterocyclic ring found in alkaloids results from there Mannich type of condensation from aldehyde and amine groups. The chemical nature of the nitrogen rings gives ways through which alkaloids can be further classified for instance glycol alkaloids are glycosylated with a carbon molecule.

The result from reactions of metabolism. Insects and herbivores initially repelled by the toxic and bitter taste of alkaloids. They are considered toxic by how they affect the nervous system due to their disruption on how negatively the effects of electrochemical transmission in the brain. For example, consuming high tropane alkaloids can result in an increase in heartbeat rate can lead to one becoming paralyzed and in extreme cases, it can lead to death. Alkaloids also have psychological effects for example glycoalkaloids are known for inhibiting cholinesterase and result in symptoms of neurological disorders. Its toxicity can be observed when they disrupt the cell membranes of the gut. The Red Clover, *Trifolium pretense* can be consumed but has high amounts of alkaloids especially in the fall season and should not be consumed by pregnant women. *Malva sylvestris* from the family Malvaceae have their flowers consumed: made of alkaloids, saponins and tannins and are known to help in the reduction of coronary heart diseases and also as antioxidants, anticancer and betters vision (Angier, 2001).

**Figure 2.7.** The Red Clover, *Trifolium pretense.*

*Source:* https://morningchores.com/edible-wild-plants/

*Protease inhibitors:* The distribution of protease inhibitors is wide and can be seen in various plants including seeds of both cultivated and wild legumes and cereals. Protease inhibitors are mostly found in many plants. Protease inhibitors are capable of inhibiting proteolytic enzyme activity within the gut of animals. They are protein in nature and can easily be destroyed by heat processes but there can be remaining activity especially, for products that are produced for commercial purposes. The activities of these inhibitors are connected to inhibiting growth and hypertrophy of your pancreas. The benefits of protease inhibitors are not really clear but in some cases, they have been reported to reduce occurrences of pancreatic cancer especially for populations that consume soybean.

Protease inhibitors have been connected with cancer of the pancreas in the set related to animals but at the same time, it is observed to also have some anti-carcinogenic effects. The Bowman-Birk inhibitors which come from the soybean have also been proven to prevent cancers that result from chemicals such as liver, lung, colon and esophageal cancer. Trypsin and chymotrypsin inhibitors occur in the seeds of legumes that are raw. Trypsin inhibitors are known for inhibiting the action of guy enzymes hence are known for their prevention when it comes to the digestion of protein and is found in many wild plant species (Angier, 2001).

*Lectins:* Lectin is derived from the Latin word legere. Lectins are able to bind with carbohydrates. Currently, proteins that are able to agglutinate red blood cells with identified specifics of sugar levels are called lectins. Hemagglutinins is a word used when sugar levels are not known. Lectins and hemagglutinins are proteins are the glycoproteins that contain at least one and catalytic domain which shows binding that cannot be reversed in certain monosaccharides and oligosaccharides. These compounds bind to carbohydrates moieties on erythrocytes surfaces and agglutinate erythrocytes without changing the nature of the carbohydrates. Lectins are glycoproteins found in many legumes and in plants with oil seeds and have the affinity for certain sugar molecules and are capable of combining with carbohydrate membrane receptors.

Lectins are able to bind with the walls of the intestines, interact with enterocytes and also affect how absorption and transportation occur during digestion leading to epithelial lesions in the case of the intestines. Despite that lessons have been found to be labile, how stable they are will vary depending on the plant species. Most of the time lectins cannot be inactivated by dry heat and usually require moisture should be present for them to be completely destroyed. Researchers have also put their focus on understanding lectins and how they can be applied in the field of agriculture and medicine. These proteins have specific characteristics that exist in the various fields of biology as more substances are being derived and as people try to understand their roles in nature. As time goes on they continue to occupy a significant place in agriculture and therapeutic fields of research.

Lectins are found in edible wild mushrooms like *Boletus edulis* and *Lactarius deliciosus*. The molecular weights of many extracted lectins from edible mushrooms fall between the ranges from 12-kDa to 68-kDa. Lectins extracted from certain edible mushrooms are non-glycoproteins such as lectins from *Tricholoma mongolicum*, *Laetiporus sulfueus*, *Lactarius deterrimus* and *Volvariella volvacea.* Other wild edible plants with lectin include the poppy plant, sunflower seeds, quinoa and nightshade vegetables.

*Saponins:* Saponins are secondary compounds that are mostly known for being non-volatile, as compounds that are active on surfaces and also their wide distribution a wild ecosystems coming from plants. The word saponins is derived from the Latin word sapo which means soap. This is because the molecules formed by saponin are similar to those of soap especially when combined with water. These molecules define the structure and are commonly referred to as triterpene and steroid glycosides. They are made up of nonpolar aglycones combined with a single or more than one molecule of monosaccharides. The soap- like nature in aqueous solutions can be explained by combining polar and non-polar structure in the molecules. Saponins are structurally complex and this affects their nature, for example, their sweet nature and how bitter they are, foaming and properties of emulsification, properties related to pharmacology, properties related to hemolysis and in addition to this their antimicrobial, insecticide activities and the ability to kill mollusks (Department of the Army, 2009).

Saponins can be used in different situations in the food industry such as the production of beverages and sweets as well as cosmetics and pharmaceutical products. The existence of the lipid-soluble aglycones and a sugar chain that dissolves in water, they are surface- active compounds with detergents, wetting, emulsification and properties that lead to the formation of foam. The Horse Chestnut, *Aesculus hippocastanum* is found in Southern Europe. Their seeds are high in saponins and foam even in the cold and show a preference for soft water when combined with soap. The seeds are a rich source of edible starch upon the removal of saponins leeching in water and then by thorough cooking. The wild yam (*Dioscorea villosa*) has gained interest in the field of medicine due to its high levels of steroidal saponins and diosgenin, a type of phytoestrogen that is used to produce oral contraceptives and originated from the the United States. Other plants rich in saponins are the soybeans and the soap bark tree which is utilized as a food additive. Saponins are also found in herbs like ginseng, *Tribulus terrestris*, *Jiaogulan,bupleurum* root, Osha, and *Collinsonia*.

## 2.5 HANDLING ANTI-NUTRITIVE COMPONENTS

Anti-nutritional factors in plant foods are known for their roles in affecting other nutrients and micronutrients which are capable of interfering with the functioning of some organs. Many of the international factors exist in foods of plant origin. The presence of Cyanogenic glycosides, alkaloids, tannins, lectins and saponins in plant foods can lead to effects that are not desired in human beings especially when they are in excess. Specific harmful insects may also result from the breakdown of these products.

Anti-nutritional factors as well as the products that result from the breaking down can also benefit the health of humans when consumed in small amounts. The mechanisms of both the beneficial and harmful effects of anti-nutritional factors occur in the same way. There are a couple of health risks factors associated with anti-nutritional factors, for example, when individuals do not have adequate knowledge about the levels that can be tolerated for certain compounds by the human body, there's also a lack of information on the level of variation of specific risks and there's also little information related to how environmental factors influence how well or the capacity of the human body to detoxify these anti-nutritional factors (Sundriyal, M., & Sundriyal, D. C., 2001).

There are several methods that can be used in the reduction of anti-nutritional factors present in wild plants, for instance, exposure to heat. Cooking and blanching can be useful when it comes to removing anti-nutritional factors and sometimes lead to the destruction of the plant cell wall which is followed by soluble compounds being reached out in the medium used for blanching. Phytic acid and oxalates can be reduced in an effective manner by different cooking and methods of blanching. These methods can however lead to the loss of important nutrients. Research was done on the leaves of the *Amaranthus* species specifically the is *Amaranthus tricolor*, *Chenopodium album*, *Trigonella foenum grecum* and *Spinacia oleracea* related to the reduction of anti-nutritional factors stated that the sole use of blanching for a period of ten to fifteen minutes was able to reduce significant amounts of phytic acid. The amounts of oxalic acid content were reduced by blanching and cooking while plants being dried and stored were not able to affect anti-nutrients especially, in the case of leaves of plant foods.

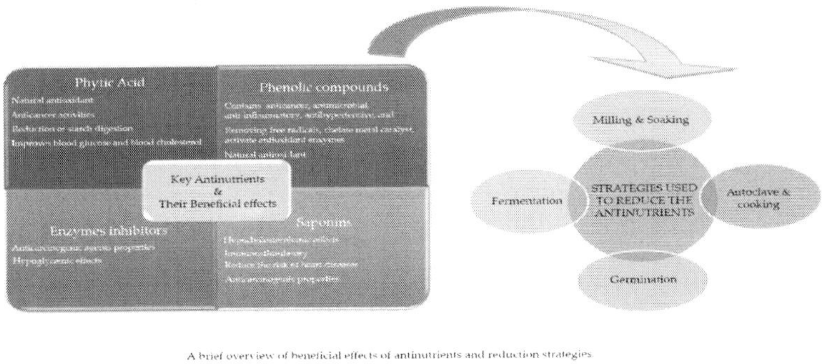


**Figure 2.8.** The benefits and the effects of anti-nutrients and their reduction strategies.

*Source:* https://fppn.biomedcentral.com/articles/10.1186/s43014-020-0020-5

## 2.5.1 Milling

Milling is an ancient method used in the separation of bran layers from grains by forming flour.it can be used in the reduction of anti-nutrients such as phytic acid lectins and tannins that are present in grains. It is a strategy that has worked to alter the chemical compounds that exist in many cereals. The chemical composition is further altered when the ground cereals are further exposed to heat. In such cases, there is a continued reduction in phytic acid and phenolic compounds and this also aids how starch and proteins are digested.

## 2.5.2 Soaking

Soaking is a strategy in the reduction of anti-nutrients and has an additional benefit that is the reduction in the time taken to cook food. When a plant food is soaked, the release of enzymes is enhanced for example the production of endogenous phytates which break down phytates. Several anti-nutrients are also soluble in water. The reduction of phytates for example after soaking will be reduced by heat treatment.

## 2.5.3 Autoclaving and Cooking

Autoclaving is the general use of heat. When this method is used the phytase enzyme is activated and acidity is also increased. When foods are boiled, especially cereals, there is a reduction in the anti-nutrient content which further improves the nutritive value of the plant food. Soaking and heat exposure also reduces the level of phytates in legume grains. Other the anti-nutrients that are reduced when food is exposed to heat include tannins and trypsin inhibitors. When seeds are exposed to heat anti-nutrients are also reduced especially, if they are soaked prior to boiling. Research has also shown that the nutritional value of legumes also improves after cooking due to a reduction in saponins and lectins. Studies have also come to the conclusion that autoclaving is one of the suitable and effective methods that leads to a significant reduction of anti-nutritive substances.

### 2.5.4 Germination

For plant foods, germination can be considered an anti-nutrient reduction strategy. It is known for its activation of the enzyme phytase which degrades phytic acid. It also alters the nutrient composition, biological properties and the physical characteristics of food. This method is commonly applied to cereals. Germination in some studies has been proven to reduce anti-nutrients like tannins and this goes on to increase the availability of some minerals also increasing the nutritive value of food plants.

### 2.5.5 Fermentation

Fermentation is an effective anti-nutrient reduction strategy when it comes to foods contaminated by bacteria. It can generally be defined as a process by which sugars that exist in plants undergo oxidation leading to the production of energy. Fermentation and enhances mineral absorption. Complexes formed by phytates are broken down by fermentation and this sets free minerals like iron, zinc and calcium increasing their bioavailability. Fermentation can also decrease anti-nutrients like protease inhibitors, phytates and tannins.

# 3

# *Classification of Wild Edible Plants*

## CONTENTS

## 3.1 INTRODUCTION

There are several theories that explain possible origin of earth. Some may argue that the world was a result of dust recollection while others may claim it began from an aspect of creation. All of these theories may correlate in describing that the world was initially inhabited by living organisms. Living organisms are entities of physical features that breathe, feed and undergo the process of growth. For instance, the bible acknowledges that planet earth was created by God. The bible goes on stating that God created animals and plants to constitute the planet`s living organisms. In his creation plants were made both domestic and wild. He further created man and allowed him to feed upon the earth's flora (Dan Dourson & Judy Dourson, 2019).

Through time to date, plants have constituted a lot in feeding the human population. However, men came up with ways of subdividing plants into both wild and cultivated varieties. Cultivated varieties of plants refer to plants that have been tamed by man. They represent plants that are grown in fields and farms that are plowed by man. They arise and develop as a result of human actions. Indeed, cultivated plants cover a vast majority of plants that are consumed by man. The majority of cultivated plants are non-poisonous as they are already researched on in modern laboratories.

On the other hand, wild plants are plants that grow spontaneously by themselves, maintaining their own growth and populations in both wild and semi-wild ecosystems. Wild plants exist independently without any human interaction. Wild plants can be interrupted by wild animals who may be beneficial to their growth. The wild ecosystems with wild plants include forests and wild bushes. Forests contribute to around 90% of wild plant growth. Most countries have vast regions covered by forests. Wild forests are the biggest pillars of climate conservancy. At the beginning of the year 2020, children all over Europe and the United States went on to protest on the destruction of forest and wild flora. The destruction of wild flora would go a long way in reducing animal food and further increase global warmth (Dan Dourson & Judy Dourson, 2019).

Wild edible plants represent plants that can be consumed by human beings. These are neither cultivated nor domesticated. Wild plants were largely edible in the initial ethnic communities that were known to live in the forests. As a matter of fact, some communities still live and depend totally on forests. For instance, the kuduk community in Kyrgyzstan represents a group of people who live in the forests. They speak their own language called kuduk and are entirely dependent on 'torang'. Torang is a term in their language that describes the jungle. It implies that the kuduk community depends on wild plants and animals for their survival. They feed on wild berries, leaves and other botanical aspects that are discussed in this chapter.

**Figure 3.1:** The Amazon rainforest; the world`s largest wild rainforest. It is home to a variety of wild edible plants.

*Source:https://www.popularmechanics.com/science/environment/a28910396/ amazon-rainforest-importance/*

Wild edible plants represent a subject in ethnobotany. It covers the vast application of plants by people in obtaining food. Research is conducted by ethnobotanists to find out the plants that are used as food and those that have already undergone extinction. Wild edible plants are further classified into various types. They include;

- Wild edible culinary fruits.
- Wild edible culinary herbs and spices
- Wild edible culinary nuts.
- Wild edible cacti
- Wild edible flowers
- Wild edible seeds
- Wild edible leaf vegetables
- Wild edible roots

## 3.2 WILD EDIBLE CULINARY FRUITS

Fruit can be defined as the fleshy product of a tree that comes as a result of plant fertilization. In biology, they represent the ovary of a flower after the process of fertilization. The ovary encloses the seeds hence the reason why fruits are found to contain seeds. Every angisospermic plant set flowers and after fertilization sets fruit. Wild plants may bear fruits that are poisonous and harmful to the health of a human being. They also constitute food for other animals that dwell in the wild regions. Wild fruits can be found in vast forest land. They tend to grow in similar habitats. This implies that plants of the same genus or species that produce edible fruits may grow within a certain radius from each other (Blair, 2014).

When you come across a plant producing fruit that is edible, it is more likely that there are similar edibles nearby. One reason for this as the agents used for dispersal. Such plants are highly reproduced by means of seeds. For example, an animal may feed on the fruit and upon motion, the same animal would move for a short distance before defecating and launching the growth of another similar plant. In another case, the plant bearing fruit may require soils containing nutrients suited for that region. Plants of the same genus will probably have similar needs and grow around the same region. This has enabled researchers to easily identify plant-bearing edible fruits in the wild ecosystem.

Fruits are closely related to berries. As a matter of fact, berries are generally fruits that resemble a juicy pulp that surrounds a seed or a number of seeds. Berries are known to grow in bulk interconnected in a similar branch while some few may grow independently as a single fruit. Those that are made up of a bunch lot of little fruits re known as aggregate fruits. Examples of berries include the blackberry, raspberry and many more. In this chapter, fruits will be intertwined with berries as berries are fruit-related.

The term culinary fruits is used to define the types of fruits associated with cooking. However, cooking is a list of processes that not only incline on the art of burning but also various processes up to and including food serving. Some fruits may be added to cooked samples of food during serving and can also be defined as culinary. Culinary set of fruits can also be used as raw products in the manufacture of other food products such as fruit jams which are then attached to cooked food during preparation and also serving. Some other sweet varieties can be used to manufacture fruit juice in which can be attached to foods of a particular cuisine. Traditional food preparation such as the Chinese cuisine readily attached fruits in their diets.

The Jewish cuisine could at times use grapes in the manufacture of wine or readily attached other special religious fruits in their cuisines. Indeed, culinary fruits hit both sides of the coin, wild and cultivated. Any form of fruit attached to cooking will bear the name culinary. In this chapter, we dive into understanding the various types of wild fruits that were attached to food during the process of cooking (Blair, 2014).

Examples of wild edible fruits include;

*Autumn olive:* Scientifically known as *Elaegnus umbellate*, autumn olive is a wild plant native to eastern Asia and covers the Himalayan ranges all the way eastwards towards Japan. The plant is known to grow and spread like wild fire. It is highly aggressive especially in covering and colonizing barren lands. It represents a small shrub basically like a small tree. It contains sharp thorns and bears fragrant flowers. Its flowers are generally white or yellow and contain four lobes. Its leaves contain silvery scales after they emerge. The plant is known to emerge during spring season after which its leaves turn green at the onset of summer. The autumn olive contains fruit that represents a small round drupe. When ripe, its fruits tend to turn from yellow to red with spots of silver or brown. Having a diameter of close to 5mm, the autumn olive`s fruit is juicy and sweet. Its fruits can be said to be in the form of a berry and can be eaten raw or used in the processing of jam (a substitute used in sweetening bread and/or cakes.). The fruit can also be used as culinary in its substitute for tomatoes. This is because the autumn olive contains a good amount of lycopene. They are used in the manufacture of the autumn olive ketchup. They can also be of use in the manufacture of smoothies and famous drink recipes.

**Figure 3.2:** The red berries of the autumn olive. They are used as substitutes for tomatoes and manufacture of smoothies.

*Source: https://leafyplace.com/trees-bushes-with-red-berries/*

The fruit upon ripening can be dispersed by birds. In the past, the autumn olive was used for erosion control, gardening, wildlife habitation and in shelterbelts. It is known to enrich the soil with nutrients and even boost fruit production of other fruits that grow close to it. *Blackberry:* This is a wild fruit of the genus *Rubus*. The taxonomical nature of fruits belonging to this genus is known to be inhabited by hybridization. The blackberry is majorly confused with the raspberry. However, it can be differentiated by the position of the torus upon plucking. The torus is the part of the stem i.e. receptacle that holds the fruit feeding it with required nutrients. When picking a blackberry, the torus tends to remain with the fruit while in raspberries, the torus remains on the plant itself.

The plant bears fruit in clusters and can be defined as aggregate fruits. The fruits are borne in small drupelets. The blackberry is made up of close to 400 species covering vast areas of Europe, Africa (especially in the northwestern regions of Africa), Asia and the north and South America. They bear biennial stems and have palmate compound leaves. The plant does not bear any flowers. However, its lateral buds break and produce flowering laterals composed of small leaves of up to five leaflets. They contain sharp prickles that can tear through the skin. Recent laboratory research conducted in the University of Arkansas has developed prickle-free forms of the blackberry plant. The flowering laterals are pollinated by insects such as bees.

The blackberry is used in the production of jam which is then used for culinary purposes during bread/cake sweetening. Bread are majorly toasted having the blackberry jam. The blackberry is also used in the manufacture of wines band in the production of sweet candy. Blackberries can easily be identified and even directly consumed by adventurers.

**Figure 3.3:** The blackberry fruit. Its transformation from a green to a ripe red is shown.

*Source: https://commons.wikimedia.org/wiki/File:Ripe,_ripening,_and_green_blackberries.jpg*

*Mountain ash/rowan berry:* This is a plant belonging to the genus *Fraxinus* and sub-genus *Sorbus.*It grows to have compound leaves and unfused fruit carpels. They are native to the Himalayan regions and parts of western China. It bears fruits that are mildly bitter and to some extent sweet. This implies that the mountain ash is not suitable for direct consumption. The fruit can instead be used for culinary purposes. Rowan berries were used during cooking as an additive in ancient Europe.

The rowan berries are not inclined onto a branch but instead hang on the main stem of the tree. They appear during winter especially after the leaves have been shed. Winter seasons are marked by a lot of snowy periods. Upon the fall of snow, mounds of snow tend to fall on the fruits making the tree look bulgy.

**Figure 3.4:** The rowan berries, it was used as a food additive in ancient Europe.

*Source: https://www.healthbenefitstimes.com/rowan-berry/*

*Oregon grape:* This plant is scientifically known as *Mahonia aquifolium.* This is a wild fruity plant inhabiting northwestern America. It is known to be evergreen and extends to a height of about 9 feet. It contains leaves that are made up of spiny leaflets. Its stems are thick and fluffy. Its yellow flowers upon pollination bear blueberries hence the name, Oregon grape.

The Oregon grape majorly grows in large forests such as the Douglas fir forest and in vast areas of the Rockies and cascades. The fruit was used for culinary purposes in the traditional communities of the ancient pacific. Such communities include the Pacific Northwest aboriginal people. They included the fruits in their diets and even used them as a mixture with a traditional entity called `salal`. Today, the Oregon grape is used in the making of wine and can be consumed raw, when ripe. It can be used to make the Oregon grape curd, which is used as culinary in the manufacture of sauce. Moreover, the berries can be used as forms of medicine to cure indigestion.

**Figure 3.5:** The Oregon grape. It was used in the making of `salal`.

*Source: https://www.naturehills.com/oregon-grape-holly*

*Partridgeberry*: It is scientifically known as *Mitchella repens.* It was initially defined as the squaw vine. It also belongs to the genus Mitchella and a member of the madder family. The wild plant represents a vine which is non-climbing. It reaches a height of 30cm in maximum length. *Mitchella repens* is evergreen with a pale yellow midrib. Its nodes are inhabited by adventitious roots. The roots may branch and spread continuously hence producing a ground root-mat like feature. The flowers of partridgeberry occur in pairs. The flowers contain one pistil, four white petals and four male stamens. The plant is a member of the distylous taxa. Distylous taxon represent a family of plants that either has long pistils and short stamens or have long stamens and short pistils. The plant species hence undergo heteromorphic self-incompatibility.

**Figure 3.6:** The partridgeberry; its fruits are drupes.

*Source: https://www.healthbenefitstimes.com/partridge-berry/*

The partridgeberry is considered special in that it takes the effort of twin flower ovaries to fuse making a single berry. The fruit of a partridgeberry is a drupe and is known to contain eight seeds. The plant is dispersed by birds who feed on the fruit. Foxes are known to consume the fruits and may disperse the seeds through their waste. The wild plant was used for culinary purposes in the making of pancake-like bread upon the combination with cornflour. Their taste when consumed raw is mild and barely sweet.

*Prickly pear cactus fruit:* This is a wild plant and member of the cactus family. However, it is a fruit- bearing cactus that falls under wild edible and culinary fruit. It belongs to the genus Opuntia. It grows in arid or semi-arid regions and bear spines on their barks. The plant contains glochids (hair-like prickles). Upon maturity, the glochids eventually detach from the plant. The plant also produces flowers which upon pollination produces fruit. The fruit`s placentation is termed parietal while its seeds are arrilate. The fruit is also thorny and should be peeled before consumption. Its prickles and glochids can be ingested causing discomfort and injuries associated with the lips, tongue and throat. Ancient Native Americans would roll the fruit in a sand medium to remove the glochoids. The fruit can also be rolled in a flame to remove the glochids. The fruit is used for culinary purposes in the cooking of soups, bread, vegetable dishes, desserts and making salads. Its stem is consumed raw but lies in the edible parts of the cacti family. The prickly pear was initially used in ancient cuisines such as Mexican cuisine. It is currently used as a substitute for corn and hence used in the making of corn-related products. This plant formulates both edible fruit and cacti. The fruits can also be strained and cooked to make the prickly pear jelly.

**Figure 3.7:** The fruit bore by the prickly pear cactus. It is also an edible wild cactus but bears fruit.

*Source: https://www.potentash.com/2018/01/23/eating-cactus-combating-food-insecurity-kenya/*

*Solomon`s plume:* It is also known as the false Solomon`s seal. Its scientific name is *Maianthemum racemosum.* the plant`s name goes a long way in depriving the plant its true status . Many think of it as a poisonous fruit, but instead, it is a cool and healthy wild culinary fruit-bearing plant. The plant bears its roots to North America. It has grown in the wild vastly across all provinces of Canada and all states of the United States of America. It is a perennial plant that grows in the woodland. The plant is basically a tree, growing to a height of 30 inches. It is inhabited by oblong-lanceolate leaves. The tree is flower -bearing hence upon fertilization leads to the formulation of a fruit. After fertilization, the flowers bear green fruits which turn red upon ripening. This occurs during the summer period. The tree would grow most especially, in riparian corridors.

**Figure 3.8:** Solomon's plume; it produces white flowers during the spring period.

*Source: https://www.finegardening.com/plant/solomons-plume-smilacina-racemosa*

During spring, the tree is depicted to be a forest bride. It produces a wide plume of white flowers. It would later convert into a red-speckled fruity tree in summer periods. The taste of the fruit is regarded as sweet and a little bit warm. One disadvantage of Solomon`s plume is that it contains a sweet flesh that is surrounded by an enormous white seed. This makes the tree look less productive. They are used for culinary purposes in the making of sauce or food jelly.

*Chokecherry:* This plant is scientifically known as the *Prunus virginiana*. It is widely known as the bitter berry or the bird berry. It was regarded as one of the most used plants in native America. The plant fruits were used in traditional cuisines and even served on major tables. The fruit bore by chokecherry is bitter (from the name bitter berry). This eventually fades away when the fruit becomes dead ripe. Chokeberries contain a small amount of flesh that covers a raw and very large pit. The pit is said to contain plant toxins that may be harmful to human health.

**Figure 3.9:** The chokeberry; they have a very large pit with a small amount of flesh.

*Source: https://www.fast-growing-trees.com/products/canadian-red-chokeberry*

In the ancient culture, the berries would have been pounded into a paste and dried in the scorchy sun. Sun-drying was a clear way of eliminating toxins. The dried patties were then used for culinary purposes and used in pemmicans i.e. meat jerky patties or a pudding dish is famously known as wojapi. Today, chokeberries are used extensively even in the making of wine.

*Abiu:* This wild fruit is scientifically known as *Pouteria caimito*. It is native to the Amazon forest of South America. It is represented by a tall tree that grows to a height of about 33 feet with a maximum of 116 feet. It is known to tower above canopies and portray its fruits to the wild nature. It is dispersed by seeds that are normally covered inside a bright yellow- skinned fruit. Its fruit is appealing to both animals and birds. It belongs to a plant family known as Sapotaceae. Wild hunters are known to capture a good harvest and register large sums of money upon sale in the markets. This is because the tree produces up to 100 fruits in a single season or perhaps a particular year. Moreover, the tree bears fruits in every 3 months. In the ancient periods, the tree was widely known by the Amerindian community. It bears elliptical leaves. Its distinguishing feature is that its flowers are hermaphroditic. Hermaphroditic means that a single flower contains both sexes.

**Figure 3.10:** The abiu wild fruit. A single tree produces around 100 fruits in a single season.

*Source: https://allfruits.blogspot.com/p/1.html?view=classic*

The abiu fruit is edible and good to human health. It can be consumed raw although its gummy latex easily sticks on one's gum. One is advised to select dead ripe samples and using a utensil such as a spoon to remove the inner content. Pulp extracted from the abiu fruit is readily used in the flavoring of ice cream and yogurt. This influences the taste of the fruit and increases calcium and vitamin A and C in the extract.

With time, the growth of the tree has been spread to forests and regions that experience tropical climatic conditions.

*The African breadfruit:* The African breadfruit is borne by a fruit scientifically known as *Treculia africana*. The tree is native to the African continent most especially in the west and central parts of Africa. The tree is a towering plant that grows to a height of 30 meters. The bark of the plant is normally grey and produces some sort of creamy latex. The tree species are normally dioecious. This is a term used to describe trees that have different sexes on different trees. However, some species which are regarded to have undergone crafting are monoecious. The fruit i.e. the African breadfruit is big, round and has a greenish -yellow color. The edible part of the fruit is mostly the seed. The seeds obtained from this tree are dicotyledonous. The fruits are very hard and are known to be taken by the chimpanzees of the forests. To explain how hard, it is, chimpanzees use specific types of tools such as stones and hard sticks to break open the fruit.

In Africa, the Nigerian community directly consumes the fruit raw. It is often used up as part of the main dish. The fruits of this fruit are highly considered. However, seeds are later on discussed in this chapter. The African breadfruit is known to contain lots of fermentable sugars and can be used in the manufacture of local brew or factory-made alcohols. In Tanzania and regions of Malawi, the fruit is used to make animal feed such as bran and the fruit head pulp.

*African mangosteen:* It is scientifically known as *Garcinia livingstonei.* It is native to the tropical parts of Africa over the vast regions of the Ivory Coast to Somalia and to the southern parts of Africa. The tree represents an evergreen plant that grows to a maximum height of 18 meters. The flowers are borne on the stem of the plant and occur in clusters. It bears a sweet and relatively acidic fruit that has a bright orange color. However, the African mangosteen contains latex that may be unfavorable during consumption. Similar to the African breadfruit, the African mangosteen is dioecious and requires different trees containing different sexes to aid in the process of mating. However, modernity has brought about grafting which allows the combination of two sexes to produce a monoecious tree.

The fruit is used as a culinary in the form of a traditional African meal. However, the fruit was majorly grown for ornamental purposes but could also be eaten. It can also be used in the production of juice. Its juice easily stains teeth while the fruit`s latex enables it to stick to one's mouth hence an agent of being uncomfortable.

**Figure 3.11:** *Garcinia livingstonei*, it grows over the tropical parts of Africa.

*Source: http://phytoimages.siu.edu/imgs/paraman1/r/Clusiaceae_Garcinia_livingstonei_126688.html*

*Bacupari:* It is scientifically known as *Garcinia gardneriana.* It is a plant species native to the Amazon region of South America. The tree bearing the fruit is an evergreen plant. It represents a dioecious species of trees. The tree-bearing fruit is however said to be of very slow growth. It grows to a maximum height of 10 meters. Due to their short nature, this tree tends to grow in the dense shades of the amazon rainforest. Traditionally, the fruit was popular among communities living around the Amazon forest. The fruit borne from this tree contains three large seeds making the thin pulp the only edible part of the fruit. It is indeed very sweet and tasty. To date, the Bacupari is rarely found due to its slow growth and minimal chances of survival. However, it is highly known and consumed raw or as a culinary additive in various modern cuisines.

*Bat-leaf passion flower:* It is scientifically known as *Passiflora coriacea.* It is botanically a vine that is fast growing. Its leaves resemble a bat`s wing hence the name bat-leaf passionflower. The leaves are ever dark-green containing spots of light green. The vine bears white-yellow flowers which eventually undergo fertilization to produce purple fruits that are rounded in shape. The vine grows to a maximum of 30 feet. The vine is mostly distributed by animals mostly birds. The vine is mainly ornamental but its wild fruits are considered tasty.

*Biriba:* Its scientific name is *Rollinia deliciosa.* It belongs to a family known as the custard apple family. It grows majorly in the Amazon region of South America. It bears a fruit known as the biriba or wild-sugar apple. The tree bearing the biriba fruit is highly tolerant of massive floods. The Amazon River is known to break its banks and spread into the thick Amazon forest. Such trees are suited to withstand such conditions. It has long leaves that are feathery and suited to withstand high water conditions. The leaves range to a height of 30 meters. The tree bears fruit in the shape of a cone. It is very large and green when immature. Upon ripening, the fruit turns yellow. The fruit is high compared to a pineapple fruit. The surface of the fruit has short spines that can easily spike and bruise human beings. The tree over recent times has gained popularity in individual human gardens. Its delicacy is slowly but smoothly increasing its popularity. The pulp extracted from the fruit is indeed very tasty and takes the taste of dessert pie famously known as the lemon meringue pie.

**Figure 3.12:** The biriba fruit. It resembles a pineapple fruit.

*Source: https://www.facebook.com/WaimeaValley/posts/rollinia-deliciosa-also-known-as-biriba-is-native-to-tropical-south-america-rela/10153908912953199/*

Most chefs especially in South America have used the fruit for culinary purposes. It is used to induce sweet flavors to commercial dishes. It has gained its way to the wine platform. The biriba is used in the manufacture of wine in Brazil and neighboring countries of South America. It traditionally depicts that the tree was an important component during the ancient war. The berimbau, (a bow of the Afro-Brazilian instrument) was sculpted out of it. During this, the warriors and young men who performed the task would pluck and feed on the tasty fruit. Sailors of the Amazon River would also use the tree to construct a boat out of its durable composure. They would also not be left out in munching the sweet fruit.

*Guava:* Its bearing tree is scientifically known as *Psidium guajava*. It grows mainly in tropical and subtropical regions including forests associated with such conditions. In some countries, guavas have become a major export commodity. Guavas were an ancient delicacy only found in the wilderness. However, its history depicts that people would chew them and either defecate or spit out. With these communities started realizing the importance of having guava trees. Today, India is one of the largest producers of guavas in the international market. In the year 2016 for example, India was recognized as the largest producer of guavas in the world. They produced 41% of the world's total guava production. There are various types of guavas the famous one being the apple guava. Guavas belong to Myrtoideae i.e. they represent a series of dicotyledonous plants. They have simple, ovate leaves with white flowers that are inhabited by a large no of stamens. The flower contains too many ovules and ends up bearing fruits containing many seeds. The fruit is normally green after fertilization and ends up turning yellow upon ripening. The seeds vary in number and hardness.

**Figure 3.13:** Guavas, they have been explored commercially in Asian countries.

*Source: https://www.healthline.com/nutrition/guava-pregnancy*

The fruit is used for culinary purposes in the Mexican cuisines and communities living in Latin America. It is used in the creation of punch, a guava-based fruit juice that is used in either hot or cold sauce in the Mexican cuisines. The guava fruit is also used in the making of desserts, snacks, candies and ales. In most third world countries, guava is taken raw and considered a delicacy among children. In Africa for example, it is very common to find young children on top of a guava tree enjoying the sweet delicacy. Some may consider adding some bits of pepper and a pinch of salt to make the fruit tastier. In a Philippine meal known as sinigang, guava is considered a major ingredient. Guavas are also a delicacy among animals such as birds. Birds are the chief agents of seed-dispersal in the guava family.

Guava is also a major constituent of pectin. Pectin is used in the large scale manufacture of jam and marmalades. This explains why major commercial industries may keep a close eye on guava farmers and more extensively on individuals who harvest wild guavas from the forests. In Brazil, industries may also lookup for guavas for the manufacture of de-goiabeira, a famous drink formed from the fusion of guavas and its leaves. The drink is used for medicinal purposes. To sum up, ripe red guavas are used as a substitute for tomatoes. This is a medicinal push especially in the minimization of acidity found most especially on tomatoes.

*Mundu:* It is scientifically known as *Garcinia dulcis*. It is a tropical tree that survives under both tropical and sub-tropical climatic conditions. Towering up to 15 meters, *Garcinia dulcis* is a complete evergreen plant that appears dense and straight. In the heavy rainforests, the mundu is known to cripple its way and survive on the harsh shady conditions. The tree has two seasons and bears flowers twice in every year. However, the tree can also grow and survive in dry conditions.

Its fruits are normally green but turn orange upon ripening. The fruit can be consumed raw. It bears a juicy pulp that contains a sour taste. Upon preservation, the fruit can be used to manufacture jam used for bread. The fruit is also known for its medicinal value in the treatment of scurvy and other basic wounds.

*Quince:* It is known as the only member of the plant genus *Cydonia.* It belongs to the family Rosaceae and is scientifically known as *Cyodonioa oblonga.* Its fruit has close similarities to that of a normal pear. It is a very attractive plant represented by its very pink flowers. The tree bearing the quince fruit towers to a height of 8 meters and bears a fruit 5 inches long. Upon fertilization, the tree bears a green fruit inhabited by grey -white hair. Due to environmental conditions, the fruit eventually sheds away its hair. Upon ripening, the fruit turn yellow the tree bears fruit which might be poisonous to the body.

**Figure 3.14:** The quince fruit. It resembles a cultivated pear fruit.

*Source: https://www.pritikin.com/quince-fruit-fall-food*

The seeds of the fruit contain a very strong compound known as nitriles. The compound nitriles chemically contain a compound called hydrogen cyanide. Hydrogen cyanide is regarded as an extremely dangerous compound whose fatality is death. Hydrogen cyanide is produced when the compound bitriles is hydrolyzed by enzymes in the stomach. Quincies represent wild fruits which are harvested for money value. In countries such as turkey and china, quincies largely boost up the countries economy. China and Turkey are the largest producers of quincies producing up to 700000 tons per year.

Cultural beliefs have it that quinces represent the fruits that led man to suffer in the current world. It is believed that quincies were located in the middle of the Garden of Eden where the almighty had directed Adam and eve not to pluck fruit from.

Quincies are used for culinary purposes in the flavoring of food. Quinces improve the food`s aroma. Quinces are too hard to be eaten raw. Due to this, the fruit is blotted by the use of frost or left to decay before being consumed. The fruit itself can be cooked and used in the manufacture of jams and jellies.

## 3.3 WILD CULINARY HERBS AND SPICES

Herbs and spices are very essential in the diet. They provide nutrients minerals and important compounds that are needed by the body. Herbs come from the leaf part of the plant while spices come from the roots, barks and seeds. They are both used for seasoning and the distinguishing factor is that any other part apart from the leaves that are used for seasoning is referred to as a spice. Herbs are essentially used in small amounts as they are used to add flavor but not substance to food (Press, 2013).

*Wild garlic and onion:* Scientifically called *Allium ursinum.* Its common names include bear leek, bears garlic, ramsoms, buckrams etc. they grow to heights of between 40-50 cm. leaves appear in March. They are best when still young and tender. They are majorly found in Britain and Asia. It grows in moist woodlands. They majorly during two seasons i.e. spring and fall. They are first dug from the ground; they are then chopped. They are then dried. The tops of the onions are chopped off. They are then on top of baked potatoes or sometimes added on top of salads. They contain vitamin A and C, calcium iron phosphorus, copper and sodium. They are also known to contain anti-bacterial, anti-biotic and traces of anti-viral compounds (Press, 2013).

**Figure 3.15:** Wild garlic. Added on top of salads.

*Source: https://www.thespruceeats.com/what-is-wild-garlic-435437*

*Sumac:* Scientifically known as *Rhus coriara.* It is native to Mediterranean Europe and western parts of Asia. They prefer to grow on the river banks. The sumac has red and white fruits. The red ones are edible but the white sumac could be toxic. The red fruits are fuzzy, have an acid taste and astringent. This is mostly used in Middle Eastern foods. They are dried, then ground. They are used with salt or sometimes pepper. It is sweet when taken with pork or chicken. It is said to contain tannins, flavonoids, anthocyanins, phenolic acid, organic acids, steroids, essential oils and terpenoids.

**Figure 3.16:** Sumac. the red fruits are fuzzy, acid taste and astringent.

*Source: https://www.1faydalari.com/en/sumagin-benefits/*

*Pine tree leaves:* The pine tree leaves are needle -shaped. The pine trees including, the hemlock and fir have a high content of vitamin C. they are dried and ground. They are used to make tea and sometimes rubbed on meat when it is mixed with several spices. The pine tree leaves are plucked and one can chew the leaves directly. They also contain vitamin A which is good for eyesight, hair and skin regeneration and help in blood reproduction.

*The garlic mustard:* Scientifically known as *Alliaria petiolate.* It is native to Europe. The seeds can remain viable for several years hence its ability to remain successful. It is also a stubborn weed on the farm. Grows mostly in the east of the United States. It tastes like the normal garlic and is used in places where the normal garlic could have been used. It prefers mostly creek edges and open forests. They take almost two years to mature but the best time to harvest is its second year when the flower buds have appeared. The young stem is the most delicious part as it is sweet, spicy and succulent. The stems can be chopped and added to soups, salads, tacos and other meals to add a spicy flavor.

The stinging nettles are known to contain a lot of nutrients, minerals and vitamins. They contain a lot of iron. In many societies, it is dried. Then used in soups. In most African communities, especially in the eastern part of Africa, it is used to make stews.

*Wild mint:* Scientifically known as *Mentha arvensis.* Another common name used is the corn mint. They can grow in disturbed land, damp wood edges etc. It is native to America as it is observed that Native American Indians used it. There are many varieties of wild mint and almost all of them are edible. They are mostly used in teas. They have a strong mint flavor and a little taste of vanilla. They can also be mixed with jelly to make a nice spread. They are also used in flavorings for salads and other cooked foods. Oils can be extracted from the plant and they are used in the sweetening of sweets and some beverages.

**Figure 3.17:** Wild mint. Have a strong mint flavor.

*Source: https://shop.balkanherbs.co.uk/wild-mint-8-p.asp*

*Chicory root:* Scientific name of the plant is *Cichorium intybus.* It is commonly also referred to as inulin. It is common in north- western parts of Europe. Chicory root is a common weed. The roots are removed and washed they are then dried. Some people grind them and use them as a substitute for coffee. They can be mixed with other spices to be used in the flavoring of meat, chicken or pork. It contains manganese and vitamin B6 which is good for brain health.

*Basil:* Scientifically, it is known as *Ocimum basilium.* Also called the great basil. It is native to tropical regions i.e. from central Africa to parts of South East Asia. Basil grows mostly during the summer season. It is quite aromatic as it has a very strong smell. It is mostly dried and ground and added to soup and stews. Fresh basil is used in pesto sauces and can sometimes be added to tomato salads. It contains sodium, potassium, iron, magnesium, cobalamin, calcium; vitamins such as vitamin A, D, C, B-6 and proteins and trace amounts of fat.

**Figure 3.18:** Bay leaves. dried and used in making tea.

*Source: https://www.burlapandbarrel.com/products/laurel-bay-leaves*

*Bay leaves:* The bay leaves which are derived from the bay laurel tree. They are used in soups and stews. It has a woody taste and is used in stews of potatoes or beans. The caraway seeds are used in flavoring. They are used to make rye breads. It contains sodium, potassium, calcium, iron and magnesium. It is usually used for making tea. It also contains vitamins A and C.

*Chives:* They are scientifically known as *Allium schoenoprasum.* The chives are a member of the lily family. They have long green stems and smell that is mild and not too pungent. They are mostly mixed in salads, used also in baked potatoes and potato salads. They can also be used in soups, sauces and dressings. The dill weed is a common kitchen garden weed. They are used in salads or mixed in hot or cold soups. It can also be used in bread and sometimes spinach pies. They contain traces of iron and potassium. They are rich in vitamin C and A. they have an onion- like flavor.

*Fenugreek:* The fenugreek is an aromatic spice and used in very small quantities because of its very strong aroma. It has a bitter flavor and is used in Indian curries and as a pickling spice. The rosemary is a common herb in the tropical regions. It is a very strong herb and is used sparingly. It is mostly used in lamb and chicken seasoning. It is also used in vegetable stews to add flavor. It can also be used in tea, soups and sauce.

**Figure 3.19:** Fenugreek. It is used in small quantities as they have a very strong aroma.

*Source: https://www.diabetes.co.uk/wild-therapies/fenugreek.html*

Most herbs and spices are collected and needed in small quantities. Most people are now being encouraged to grow these herbs and spices to get compounds and nutrients good for their health as they are free growing and do not need chemicals. The compounds found in these plants are sometimes distinct to them. They act as anti-inflammatory and anti-oxidants.

## 3.4 WILD EDIBLE CULINARY NUTS

A nut is considered to be important in the human body. This is because it is rich in nutrients and provide a substantial amount of energy due to their high oil content. Most of the nuts are a source of vitamin E and $B_2$. They are a source of fiber and folate and fiber and are sources of various minerals essential to the body such as phosphorous, potassium, copper, selenium and magnesium. They are not only important to humans but also wildlife. It comprises two parts which include a hard shell and a seed. The hard shell is inedible while the seed is edible. It is sometimes difficult to understand the botanical context in relation to nuts. This is because according to this context, the nut does not open to release the seed. However, the word nut has a lot of ambiguity in translation in most communities. Unlike most seeds that wildly disengage themselves from the shell, nuts have compact shell walls that are from the compound ovary. Most nuts in the botanical context are not nuts in the culinary context. In the botanical context, a plain dry fruit with a hard ovary wall and in which the seed remains unattached with the ovary wall is a nut. Hazelnuts and chestnuts are considered as nuts in the botanical context but are also culinary nuts (Ballerini, 2012).

**Figure 3.20:** Hazelnuts, used in the manufacture of chocolates.

*Source: https://www.newfoodmagazine.com/news/76450/hazelnut-metabolites/*

- Hazelnuts- if one is looking for an ingredient used in making Cadbury chocolate bars, then this is the one. Hazelnut is used in combination with chocolate to make chocolate truffles and other products such as Nutella. The oil of these nuts is highly flavored and is used as cooking oil. The nut is edible and can be ingested raw, roasted or can be ground to make a paste. This nut contains various nutrients which include protein, fats, vitamins and minerals. When making torte, hazelnuts can be used as an ingredient. The hazelnut can also be grounded to make flour and can be used in various cuisines. The hazelnut has different flavors. They also contain a substantial amount of dietary fiber.

- Chestnuts- these nuts originated from the temperate regions of the northern hemisphere. The fruit is within a cupule and can contain one to seven nuts. Unlike other nuts, chestnuts contain very little protein and their calories are mainly from carbohydrates. They have very little fat. The nut can be ingested raw which requires peeling or can be roasted which does not necessarily require roasting. The nut can also be milled to acquire flour and thereby used in baking or can be used as a stew, sauce soup thickening. The nuts can also be grilled, steamed, deep-fried or even boiled.

In the culinary context, a nut is a dry edible seed or fruit that in most cases but not always have a high- fat content. In some cases, one can consider the oily kernels that are found within a shell and are edible are known as nuts. Humans have been able to use nuts in various ways such as; using in cooking, eating them raw, roasted to be as a snack food and in other situations they are pressed for oil that can be used while cooking or can be used as a cosmetic. Nuts though widely used as food by humans, are the most common causes of allergies thus mostly referred to as food allergens.

Fruits and seeds that are considered as nuts in the culinary context but are not considered as nuts in the botanical context include:

- Brazil nuts- is known as *Bertholletia excels.* Is a tree originally found in South America that is considered the longest- lived tree in the Amazon Rainforest? The tree has commercially edible seeds that are harvested. Are widely known for its richness in micronutrients. Due to its diverse use as a nut, the nuts are protected as it is illegal to cut down the tree. The nut can be pressed to produce oil which contains unsaturated fatty acids. Apart from cooking, the oil can also be used as a lubricant in clocks, in the manufacturing of paint and in the cosmetics industry. Due to the hardness of the Brazil nutshell, individuals often crush it to use it as an abrasive to polish metals and ceramics. The tree can also be used as wood. The tree is considered as that of very good quality, having various uses such as flooring and even heavy construction.
- Cashew nuts- the cashew tree is scientifically known as *Anacardium occidentale*. It produces the cashew apple and the cashew seed. In the culinary sense, the cashew seed is referred to as a nut in the culinary context. The nut can be eaten as is, used in recipes or processed into cheese or butter. It was originally found in the Caribbean islands, Central America, northern southern

America and northeastern Brazil. The shell of the cashew nuts can be used in many applications. They include lubricants, water-proofing, and paints and was used in arms production. These nuts are widely used by Indians in Indian and Pakistani cuisine. Are used for garnishing sweets and curries. One can also grind them into a paste that are used as sauces. In powdered form, they are used in the making of several sweets and desserts. The shell of the cashews is have oil that can cause dermatitis that is not good to the human body. Due to this, the shells are extracted for research for its application in biotechnology and nanomaterials. For some people, cashews are considered allergens.

- Peanut- are also known as groundnut. Scientifically, it is known as *Arachis hypogaea.* It is widely known as a legume crop and is edible. Is widely grown in the tropics and subtropics. Most individuals regard this nut as a grain legume and an oil crop due to its high level of oil content. Peanuts shelter symbiotic nitrogen-fixing bacteria in root nodules. In this case, peanut requires fertilizers that contain less nitrogen fertilizer. Due to this, they improve soil fertility and are valuable in crop rotation. Peanuts have a variety of uses in industry. They are rich in nutrients that are essential to the human body. They provide calories and are a very good source of several vitamins namely B and E, various minerals that are important to the body and dietary fiber. They are also rich in protein.

Some researchers have come to the conclusion that having a substantial amount of peanut in your diet reduces the risk of contracting certain diseases that is associated with high mortality. It reduces the risk of contracting heart diseases. It is therefore not at all absurd to say, peanuts a day, drives diseases away. Peanut is common cooking and salad oil. It however has cases of allergic reactions though few are associated with high mortality. It reduces the risk of contracting heart diseases. It is therefore not at all absurd to say, peanuts a day, drives diseases away. Peanut is common cooking and salad oil. It however has cases of allergic reactions though few. For those with allergic reactions, individuals should be careful in studying ingredients as most processed foods are inclusive of peanut. If peanuts are not stored correctly, it may be subject to aflatoxins that are very harmful to the human body.

- Almonds- this seed that is classified as nuts whose tree is native to the Iranians and the surrounding countries. The tree contains a fruit that consists of an outer hard shell which includes the seed. Almonds can be sold shelled or unshelled. Some almond seeds are sweet while others are bitter. The bitter flavor involves a recessive gene which makes it easier to domesticate. Bitter almonds were generally used for medical purposes but were severe even in small doses. Almonds are used as a component of various dishes but mostly eaten on its own, raw or roasted. Almonds can also be milled to be flour. There are also whole and sliced. By-products of almonds include almond oil, almond butter and almond milk. An individual can choose to use the almond products in both sweets and in savory dishes. If one wants nuts to sprinkle over breakfast, then almonds are the right ones. This goes well especially with dishes that include ice cream. Almonds are also used in making various kinds of desserts which include cakes, cookies, pastries and marzipan. For those that suffer from peanut allergies, need not to worry because instead of peanut butter, they can use almond butter. For individuals who are lactose intolerant, instead of using cow milk, they can use almond milk. It is fascinating since one can use almond milk as it is without using heat. Polyphenol is a compound present in the skin of almonds and consists of flavanols and other phenolic compounds. The compounds in conjunction with the skin dietary fiber are used commercially as food additives and supplements. Almonds are an almost complete diet. It consists of water, carbohydrates, protein and fat. It is also rich in vitamin B and E and essential minerals such as manganese, phosphorous, magnesium, copper iron, calcium and zinc. It also has a substantial amount of dietary fiber that are essential for digestion. They are also pressed to produce oil that can be used for cooking. Almonds are easily harmed by aflatoxin as it can affect them in the soul. In such a case, humans should be careful of such molds before ingesting them.

- Macadamia- these nuts have a very creamy texture and their flavor is butter-like. The origin of the macadamia nuts in Australia. These nuts are beneficial to the human body once ingested as they contain various nutrients. They are rich in calories, healthy fats, vitamins and minerals. Are low in compounds that are harmful to the human body when taken in high contents. These compounds are carbs and sugar. They also contain substantial amounts of fiber. As such, this is the food that is suitable for people with diabetes. Researchers have concluded that it is healthy for individuals and should be included in diets. The seed is within a fruit that is hard, woody, globose follicle and has a pointed apex. Inside, fruit may be one or two seeds. The nut can be pressed to produce macadamia oil that is used in cooking or as salad oil.

- Pine nut- are seeds that are from the pine tree and are edible. Other names include Pinole and pignoli. Other species of pines produce large seeds that are big enough for harvest, although most of them are so small to be of value to the human body unless one eats a very large amount. The fact about pine nuts is that not all pine trees produce nuts. Most countries include pine nuts in their cuisines. Others may even say that pine nuts are a global cuisine. The growth of the nut depends on the type of species it is. The reason as to why pine nuts are wild is because the ecosystem conditions in wild forests are very suitable with no contamination. Wild forests are also characterized by clouds and rains which are the best conditions for cone formation and development, otherwise, pine trees tend to abort cones. The best areas for the formation are topographical areas. It is very fascinating as to how the nuts develop. For instance, in America, some species grow and develop during spring. It begins when spring begins at which at this point pollination occurs. There will be a formation of a small cone in the middle spring till the conclusion of summer. As at that point it will remain dormant till the next spring when it will continue in its development. However, in topographical areas found in the low elevation areas, this ideology does not apply. Such areas contain shaded canyons that help maintain the humid temperature throughout spring hence providing suitable conditions for the development and maturity of pine cones seed production. Winter is considered the destroyer of pines as the sudden and unexpected changes in humidity and winds make them gullible to freezing and drying, which damages the pine tree permanently. Once the cone is ripe and harvested, it is dried so that seeds can be extracted, and is done in several ways. The dried cone is placed in a gunny sack and struck on a hard surface. This is the most common method of extraction. After the striking process, the seeds are isolated from the residue in the sack by hand. Instead of personally drying, an individual may decide to wait for it to happen in the tree before harvesting the cone. Pine nuts serve as food for both animals and wildlife. Overharvesting is, therefore, discouraged as pines are crucial to the ecosystem. The height of a pine tree as compared to sea level is very important when it comes to pine cone production. At places, where the pines are below sea level such as the shaded canyons, the conditions are favorable for seed production. When it comes to those above sea level, temperatures are so high that they can destroy the pine cones. It is encouraged for individuals to include pine nuts in their diet as they have various nutritional values. Once dried, pine nuts include water, carbohydrates, protein, and fat and have a substantial amount of micronutrients such as copper manganese, phosphorous, zinc, copper, vitamins E, K and B among others. Pines are in most cases added to meat, vegetable dishes, fish and salad. It is a major component in the Italian pesto sauce. While making a cake, it is used in topping the cake and one can optionally decide to sprinkle icing sugar. One can also decide to grind these nuts to make flour and use it in baking bread. In Catalonia, it is a component used in making sweets in which a marzipan ball is made and sprinkled with pine nuts. It is then immersed in beaten eggs and then is lightly cooked. The sweets are famously known as panellets. In southwest France, pine nuts are a major ingredient in making *salade landaise* , a famous salad in this region. The individuals in Nevada fancy pine nuts as they have a sweet fruity flavor and therefore widely promoted. In the Middle East, individuals understand how beneficial and sweet pine nuts can be and therefore use them in making most of their cuisines and dessert. In New Mexico in the United States, pine nuts are used in preparing beverages. It makes a dark roast coffee that has a nutty flavor and is called the pine nut coffee. Pine nuts can be roasted to make a very delicious snack. In Korea, pine nuts are milled to make their famous porridge known as Jatjuk. Even though pine nuts are considered to have a very nice taste when ingested raw it can be very unpleasant. It has been said to have a salty metallic taste that can last for a few days or even weeks for that matter. This should not be of worry to the consumers of this nut as it is not harmful to the human body and it can be taken care of without any treatment.

- Pecan- this nut as very common to the native individuals of the southern United States and northern Mexico. The pecan tree is very large and sheds its leaves during various seasons hence a deciduous tree. Its fruit is not necessarily a nut but can be categorized as a drupe containing a stone encompassed by a husk. Though considered a wild tree, in recent years, individuals have domesticated his plant. It is used for commercial purposes hence transferred to the gardens. The trees find the summer season very friendly and a huge advantage to them especially when the summers a long, humid and hot. For those that have decided to grow this wild plant in their farms, they should ensure that they grow several of them to allow for pollination. However, the growth of this plant is faced with a lot of limitations. The tree is susceptible to a wide variety of pests, diseases and physical disorders. These limitations can hinder the growth of the tree and the production of fruits. The trees are also susceptible to various infections brought about by fungi and bacteria that are very common especially in humid conditions. Insects such as ambrosia beetles, hickory shuck worm, curculio, weevils among other aphids feed on stems, leaves and the developing nuts. These limitations are the main reason pecans are grown in the gardens so that they can be taken care of by humans. The seed of the pecan tree is the edible pecan nut. The seed can be eaten raw or roasted as a snack. They are very sweet as they have a rich buttery flavor. These apples to butter lovers. These nuts are mainly used as additives to cookies, cakes salads and candies. It can be used as an ingredient in making various desserts such as the pecan pie and the praline candy. If one is looking for an appetizer, then pecans nuts are the right dish. In the southern United States, pecan nuts are used in preparing traditional dishes. One fun fact about pecans is that one can use them when he or she uses very huge amounts of energy. So for those who go to the gym and work out most of the time, pecans are the best emergency rations instead of ingesting a basic diet. Another fascinating fact is that pecan wood can be used when smoking meat and grilling hence giving met and the grilled food a very sweet nut flavor. Pecan nuts can also be pressed to produce pecan oil and can be used in making pecan butter. Like most nuts, pecan is important as it has various nutritional values. It contains water, protein, fat and carbohydrates. It also contains dietary fiber and various minerals such as phosphorous, zinc thiamine among many others. They are an average source of iron and vitamin B.

- Walnut- a nut? Some researchers deny they are not true botanical nuts. It is the edible seed of a drupe. However, it is commonly consumed as a nut. At the instance when the seed is ripe and its outer protection removed, it can be used to decorate a meal once placed on a plate or can be used as a snack. In recent years, walnuts have had the opportunity to be grown in farms for the purpose of both consumption and commercialization. Walnuts are always at high risks of fungal infections and mold development and hence they should be stored properly to prevent the formation of aflatoxins. In the markets, one can encounter two forms of walnuts. There are those that are shelled and those that are unshelled. After processing, they also appear in various ways. The seed, commonly known as the walnut meat which may be whole, halved or in small segments. Walnuts can be eaten raw or one can decide toast so as to portray their actual flavor. One can also decide to bake walnuts without adding any other ingredients. One can also decide to use walnut as an ingredient of a dish. Some of the dishes that use walnut as an ingredient include; banana cake, brownie, fudge, walnut pie and walnut soup. In most cases, walnuts are pickled or candied. In the case of pickled walnuts, it involves preserving the whole nut in which it can be spicy or sweet depending on the solution it is was used to preserve. One can also use walnut to make butter either at home or in the industry. Are also very good ice cream toppings and are a substantial ingredient in various dishes across the world from Caucasian chicken, chicken in walnut sauce to poultry or meatball stew. Walnut can also be pressed to produce walnut oil which can be used in dressing salads. The oil cannot be however used in frying. The nutritional components of walnuts that are shelled include water, protein, fat, carbohydrates, dietary fiber and dietary minerals such as manganese and B vitamin. This so far might be considered as the healthiest nut as compared to other nuts. To those individuals that are susceptible to walnuts, it might lead to allergic reactions. For those individuals that love walnuts are very lucky as the daily consumption of walnuts is not harmful but rather beneficial to their health. It can be said that walnuts have an amazing appetite control power. Due to the presence of omega-3, plant sterols and vitamins in walnuts, its ingestion can help subdue hunger. As such it prevents one from ingesting a lot of food and as such helps in weight loss. For women, walnuts are the right food. It is known to contain nourishing fats that can attract moisture from the environment hence soothing dry skin. They are said to be agents that soldier the skin against harmful UV rays. Some that have seen the positive results have nicknamed walnuts as beautiful skin superfoods.

- Acorns- are commonly known as oak nuts. The oak trees consist of cupicles that encompasses one seed. They contain a substantial amount of carbohydrates, water, fat, vitamins, protein and minerals. Humans are not the only consumers of oak nuts. A good number of wildlife depends on oak nuts. Some cultures consider acorns as their dietary staple. Over the years ,however, they have been replaced by grains. Acorns can be milled to acquire acorn flour that can be used to make bread. They are in most cases used as substitutes for coffee. In other tribes, acorns are consumed alongside huckleberries, seaweed or salmon. This is the only nut that can be stored in a very long time without going bad. To produce more acorns oak trees should be planted further away from each other as they won't be competing for sunlight.
- Beechnuts- are nuts that are borne of the beech tree. They are normally found in small burrs that are released from the tree during autumn. They contain fat and can be pressed to release oil that can be ingested.

With the case of the above nuts and generally all nuts, they should be consumed in the morning alongside breakfast. It is believed that a nut in the morning, prevents one from undergoing fatigue and helps regulate the levels of blood pressure. Researchers have concluded that instead of ingesting sugary baked products such as cakes as snacks, one should consider nuts as they are very sweet and furthermore healthy. Some may even say that nuts have grown to the name "pick-me-up" snack. Since nuts contain substantial amounts of fats, one should be careful as they can experience weight gain unless other fats are recused from their diet (Ballerini, 2012).

## 3.5 WILD EDIBLE CACTI

Deserts existed long before the modernization of human nature. In ancient culture, deserts existed both cold and hot. However, many may argue that the growth of desserts is attributed to human behavior. Currently, a fifth of the land on our planet is covered by deserts. The Sahara Desert for instance is the world's largest hot desert covering a massive area of 9,200,000 square kilometers. The desert was once a region covered in a green carpet. It yielded plants and crops for ancient African communities. Over time, the activities of man became barbaric to the Sahara land. Trees were fell, vegetation was burnt down and destroyed while no one cared how the future generation would look upon it. Today the Sahara is one of the harshest deserts in the world. In fact, it is ranked the third -largest dessert inclusive of both hot and cold desserts. The Sahara is however believed to be converting in a period of every 20000 years. This means that after a period of being dry for 20000 years, the dessert will convert into savannah and become green again. On the other phase of a similar period, the land would convert and become a dessert once again. The current activities of man have far much proved to be destroying our own ecosystem. One would imagine that our planet is warming up at a very alarming rate. Greenland ,for instance, is melting its ice at a rate that is worrying. All over, crucial connections between man and our planet are being disrupted. The stability that man's life lies upon is slowly being lost. Deserts are indeed encroaching faster than it has ever been thought (Blair, 2014).

However, some may argue that stabilizing the growth of desserts including the spread of arid and semiarid conditions on our lands, a process is known as desertification would go on a long way in changing our view on desserts. Desserts have a silver part of the coin too. They too issue man with benefits that may not have been available without them. Desserts are habitats to some communities such as the Tuareg ad Nubians of native Africa. They also use animals such as the scorpion which are used during hunting by the Brazilian communities living in the Amazon. At the helm of it, all desserts can be important in the provision of food to animals.

This chapter would focus on the wild edible cacti that thrive in desserts or other semiarid conditions. Most deserts are indeed considered wild because of their harsh conditions. There are some deserts where rain has never been recorded. During the daytime, one may experience immense heat while night may bring about a dangerous cold that threat a human`s life. Some plants however seem to cope up with these harsh conditions. As a matter of fact, to them deserts are a very vital refuge. They represent a romance of nature that keeps them going. One of these botanical factors is cacti. Cacti develop and grow on desert oases. Oases are locations on desserts that bear little amounts of water that help plants and animals survive. With this, the desert is regarded similar to the wild and hence the cacti a wild edible plant.

With this, man has researched and found samples of wild cacti that are not only edible but a delicacy that is of great importance to the human body. Today it is not only communities that live and survive in these conditions that would benefit from them, but also a number of individuals who would toss a coin to buy these nutritious products (Blair, 2014).

Cacti belong to a family known as the Cactaceae. Most varieties of edible cacti belong to a special species of dessert plants known as the Opuntia species. Some of the cacti plants belonging to this treaty include;

*The sword spear:* This is a species of cacti scientifically known as *Acanthocereus tetragonus.* Now, this variety of cacti is native to parts of central and South America including Mexico, Florida and the Caribbean. Ithas been awarded a variety of names including the dildo cactus, triangle cactus and organo-alado-de-piyata. It is one of the tallest cactus towering to a height of up to 7 meters. Its stems are dark green and have up to five angles. Similar to other various types of cacti, they contain areoles. Areoles are small bumps especially ranging from a light to dark color that appear on the stems of cacti. Areoles distinguish cacti and contains spines that protect the cacti against any form of danger or threat. Cacti necessarily don't have branches but the areoles represents sets of highly specialized branches for cacti. The sword spear bears flowers that highly distinguish themselves through the creamy white pistils and greenish-white tepals. Desserts are hot hence the flowers cannot open during the day. Instead, these plants open up during the night from midnight till dawn. They end up attracting hummingbird moths which will carry on the process of pollination. The cacti would then produce shiny red fruits.

**Figure 3.21:** The sword spear, it towers to a height of about 7 meters.

*Source: https://pixabay.com/photos/cactus-c%C3%A9r%C3%A9us-3340344/*

The young stem of this cacti plant is very edible. It can be used as a vegetable in which it can be cooked or consumed raw. The fruits from this cactus are sweet and eaten raw. Communities around arid and semiarid conditions may use this plant for ornamental purposes or use it as a fence for their camels and security purposes.

*Browningia candelaris:* This is a species of cactus that is native to parts of northern Chile and southern Peru. It grows straight with a spiny trunk with spineless thinner branches. Pastoralists serve as a major hindrance to the growth of this plant. Their livestock is more likely to destroy young seedlings that struggle to survive. This species of cactus resembles a normal tree. It grows straight to a height of around 6 meters. There, it begins to develop unbranched trunks that grow to a diameter of about 50cm. the top canopy like part is dense with straight brown spines which spike during harvest. The cactus bears white flowers which eventually undergo pollination to produce fleshy fruits that turn yellow when ripe. The fruits are highly edible and indeed, very sweet.

*The saguaro:* The cacti species is scientifically known as *Carnegiea gigantean.* The Carnegiea is a monotypic genus. The cacti are tree- like and grow to a maximum height of about 30 meters. The cacti grow widely in the Sonoran Desert of Arizona. It is in fact the state flower of Arizona. The cacti species is honored to Andrew Carnegie, who designated his life in protecting the cacti species and its desert habitat. The saguaro is one species with a very long lifespan. They say that a child may plant it but may never live to see it in its mature nature. The saguaro lives for at least 150 years. It takes a period of 75 to 100 years for the cacti to produce an arm. Having more arms is important in ensuring that the plant bears more flowers and numerous fruits.

**Figure 3.22:** The saguaro, a towering species of cactus that can live for over 150 years.

*Source: https://en.wikipedia.org/wiki/Saguaro*

In a single period of sporadic rain, the saguaro has the ability of absorbing lots of water, it in turn, uses it in small bits hence increasing its chances of survival throughout the intense drought. The rate growth of saguaros is totally dependent on precipitation levels. Where there is more rain, then the saguaro would grow for a long period of time. The longest or rather tallest saguaro ever measured was in cave creek Arizona which towered a height of 23.8 meters. A saguaro that does not bear arms is called a spear.

The saguaro has very sharp spines for its protection. One story in history depicts that a saguaro once separated a sheep from its skull after the sheep collided with the saguaro. This depicts how dangerous the saguaro spines are. Though aseptic, the cactus spines are very dangerous and may be very hard to remove. They may also lead to skin inflammation, a condition known as granuloma.

The plant produces white, waxy flowers which upon fertilization bear fruits with numerous tiny seeds. The flowers are pollinated by honey bees, bats and white doves. Pollinators can be categorized into diurnal and nocturnal pollinators. Other pollinators may perform a better job than others. For the saguaro, it is considered that bees are the best and most effective pollinator.

The fruits of the saguaro are edible together with its seeds. The fruits appear ruby red and in fresh or dried. The pulp can be manufactured into pulp and drink.

## 3.6 WILD EDIBLE FLOWERS

Flowers are mostly used for ornamental purposes. They are mostly used to send a message to someone e.g. to show love, to say sorry and many more. They are majorly seen during many functions mostly weddings and funerals and during St Valentine's Day in the month of February.

Some flowers are edible and can be used for nutritional or medicinal purposes. Although they might be edible for consumption some of them contain pollen which causes allergies to some users. Some flowers are regarded as poisonous; a person needs to know a specific flower to look for and feed on because they might consume a flower that will cause serious damage. Flowers can be used to add flavor and color to dishes at the dinner, breakfast, brunch or lunch table. They can be used in salads, beverages and sauces (Dan Dourson & Judy Dourson, 2019).

*Wild roses:* Roses are very common flowers. There are almost 150 different species of roses of different sizes and colors. These roses are oval or spear- shaped. Most of them have five petals. The younger flowers have a much darker color. They can be found in dry fields and open woodlands. They are mostly found in the northern hemisphere. They prefer a small shade, moist soils and they should also be well- drained. All roses are edible although they all don't taste the same. The more pleasant a rose - it tastes sweet as well. The edible parts of rose flowers are the petals. They are aromatic and have a sweet flavor. They can be eaten raw, mixed with different foods, salads or even dried and added to granola. The rose petals can be pounded and added to other liquids to produce beverages, jams or even jellies. They can also be dried and consumed to prevent heartburn. The rose is known to contain compounds that are relaxing and reduce anxiety.

**Figure 3.23:** Wild Rose flower. Has a sweet and aromatic flavor.

*Source: https://www.alamy.com/close-up-of-a-single-wild-rose-flower-with-leaves-calgary-alberta-canada-image278174157.html*

*Hibiscus flower:* The Hibiscus flower, scientifically known as *Hibiscus sabdariffa* grows in tropical and subtropical climates. It is native to East and West Africa and Southeast Asia. They are found in a variety of colors i.e. red, white, yellow and pink. The fresh calyx can be eaten raw but it is mostly used in tea, jams, salads and relishes. Many communities take it in form of tea because of its medicinal purposes as it helps reduce hypertension and the amount of cholesterol in the body. The floor of the tea is mostly sour and is dark hot or sometimes with ice on a hot day. It is rich n protein and citric acid. The citric and pectin are the compounds needed in making jams and jellies.

**Figure 3.24:** A red hibiscus flower. used in making tea.

*Source: https://www.123rf.com/photo_9956507_red-hibiscus-flower.html*

*Dandelion flower:* The dandelion is a common garden weed and is highly nutritious. It has bright yellow petals. It is known for its property of being antioxidant. They also used because they can be used as an analgesic i.e. as a pain reliever. The flowers can be eaten raw or put in a salad. They can be mixed in dough and eaten in the form of bread, fried and sometimes make jelly or even wine. They contain minerals such as potassium, magnesium and iron which can be used to make vinegar. When being used to make tea fresh flowers are picked and put in a jar or flask then pouring boiling water. It is then left for a few minutes to infuse then it is strained. The flowers can be added with honey, the nutritive properties of the dandelion i.e. vitamin C, beta-carotene which when mixed with honey. When making it the flower petals i.e. the yellow petals are mixed with raw unprocessed honey. The petals are put in a jar then raw honey added, then mixed with a knife. After this, the air bubbles are removed and left to sit on the shelf for some time. After a few days, it is strained by slightly heating over small heat. A mesh strainer is then used to remove the petals.

**Figure 3.25:** Dandelion flower. Eaten raw or sometimes in a salad.

*Source: https://en.wikipedia.org/wiki/Taraxacum*

*Lavender flower:* Lavender grows in the Mediterranean region mostly in northern Africa. They are violet in color. They are small in size but grow in very large numbers. It is mostly known for its fragrance that is mostly used for calming. Its color and aroma combined make it desirables as a food. It is added to a large number of foods i.e. baked food (bread), syrups, liquor, tea and in herbal concoctions. Lavender is used in very small amounts as it is very overpowering. Flowers that are full of color are the best to be harvested because the fresher and full of color the flower is the more the flavor. These flowers can be dried although if dried improperly the taste and flavor can be lost. When drying them, the stems are tied together and hung from the ceiling. They are hung upside down with the flowers at the bottom. They should be dried in room/s that are dark, free from dust and has good ventilation which helps in quick and complete drying. They are then stored in an airtight jar so as not to lose their oils.

*Nasturtium:* Nasturtium is a brightly colored i.e. bright orange, yellow, or red and very attractive flower. It is known to please most people because of its taste and smell. It is mostly used as a seasoning. They have a peppery, spicy, mild flavor that provides a soothing effect on a person. It contains a lot of minerals, and compound that act as anti-oxidants and anti-inflammatory. They can be eaten raw or cooked and sometimes they can be added to green salads. The flowers can be eaten when they are still buds and also when in full bloom.

**Figure 3.26:** Nasturtium. can be eaten raw and sometimes cooked.

*Source: https://paramountseeds.com/product/nasturtium-edible-flower/*

*Purslane:* Purslane produces flowers that are tiny and yellow. It has high nutritive value as it contains a lot of minerals and vitamins but its main selling point is it contains omega-3 fat. It can be consumed raw or in a salad. It can be steamed with other vegetables and served as a side dish or served in soup. Some people prefer it buttered or sometimes fried.

Pansies are small flowers that are approximately 5-8cm in diameter. They have a variety of colors although the most common are purple, blue or yellow. They have a mild, fresh and floral flavor. They are mostly eaten raw, due to their color they are used to decorate desserts e.g. pastries, cookies and sometimes cakes. They can also be eaten as salads. They are known to contain anti-oxidant and anti-inflammatory properties.

*Clover:* Referred to as *Trifolium spp* scientifically. Clovers are common weeds. They have edible flowers that are white or red in color. The white clover is more preferred to the red one. The whole flower can be eaten but the petals are the sweetest. Clovers should be consumed fresh and/or dried. They are mostly consumed in tea. They can also be added to soups, stews and lasagna. I contain compounds beneficial to the body; it contains high protein content, beta-carotene, vitamin C, the B vitamins, inositol, biotin and bioflavonoids.

*Marigolds;* Scientifically, they are known as *Tagetes repens.* It is native to the Aztec community. Marigolds also referred to as "Mary's gold" in reference to the Virgin Mary. They are found mostly in orange or yellow but sometimes they can be a combination of both. They grow in tropical regions. They can be served in salads, sometimes used to make tea. It can also be used to color food. It is known to contain zeaxanthin and lutein which help in the improvement of eyesight. When harvesting its flowers they are plucked from the stem, the petals are then removed. Ensure to remove all green or white parts to remain with the yellow petals.

*Borage:* Scientifically known as *Borago officinalis.* The borage or also referred to as starflower is blue, white or sometimes it is pink in color. The flowers are star- shaped and they are hanging facing downwards. It can be eaten as a salad, sometimes added in desserts and cocktails as a garnish. They can also be cooked in soups, sauces or as pasta fillings. They have a sweet flavor similar to the taste of cucumber and honey. They can sometimes be crystallized to make candy. When being harvested the flowers are plucked from the stem and the back stem removed.

*Snapdragon:* Its scientific name is the *Antirrhinum spp.* They are referred to as dragons because of their fictional face that resembles a dragon and when squeezed it opens and closes. The snapdragon flowers are in a range of colors; red, yellow, pink and white flowers. They can be bitter. They are mostly added to salads. The snapdragon in most places is used solely for ornamental purposes. The taste of the snapdragon is solely bitter but majorly depends on where it was grown. They contain nutrients and compounds that help the body i.e. vitamin C, potassium, iron and phosphorus.

*Calendula:* The calendula flower commonly referred to as pot marigold. Its scientific name is *Calendula officinalis.* They can grow in almost all types of soils. The flowers are picked continuously because if they are allowed to produce seeds they stop flowering altogether. The flowers are picked in the middle of the day when the dew has evaporated. The flowers are then dried in well- ventilated room and they should be shuffled to prevent them from sticking on each other. When being prepared to be eaten the petals are removed and these can be eaten raw or mixed with other vegetable salads. The flowers can also be dried and added in soup. They can also be used in making tea. They mostly taste like honey.

*Cornflower:* It is scientifically known as *Centaurea cyanus.* The cornflower is blue in color. They grow in Europe. They grow as weeds and prefer cultivated or disturbed ground. It has a spicy, sweet, clove-like taste. The flowers can be used as single florets or as a whole. They are used in salads, garnishes or blended to produce beverages. It contains biotin, vitamin

C, calcium, flavones and folate in the flower bud. It is sometimes used to make tea.

**Figure 3.27:** Cornflower. Spicy, sweet and tastes like clover.

*Source: https://www.growveg.com/plants/us-and-canada/how-to-grow-corn-flowers/*

*Wood sorrel:* Its scientific name is the *Oxalis stricta.* The wood sorrel flower is small in size, they contain five petals and grow to 1-1.5 cm in width... They appear similar to clover but have a distinct difference; the wood sorrel leaves are heart- shaped whereas the clover leaves are rounded. They can also be identified by their behavior where they tend to fold up in the evenings and open up in the morning. They are sour and provide a great taste in salads, seafood. They have pink, yellow or white flowers. The sour taste comes from the oxalic acid that makes it to be consumed in small quantities. They are rich in vitamins A and C. the flowers can be dried and used to make tea.

*Prickly pear:* Scientifically they are known as *Opuntia ficus indica.* They are mostly found in Mexico, South America and some parts of Canada. The flowers blossom between April and June. They grow in semi-arid conditions. The prickly pear flower is yellow, red or orange. They have a mild minty taste and preferred eaten raw. They contain nutrients such as proteins, fatty acids, linoleic acid and selenium.

There are many other flowers that are edible these include; angelica, anise hyssop, bachelor's buttons, bee balm, begonia, carnations, daisy's, phlox, scented geraniums ,etc. these flowers are similar in how most of them are prepared and eaten. These flowers also contain almost all the same minerals, compounds although might have their own unique compound.

Flowers contain potent anti-oxidant and anti-inflammatory compounds that are useful to the improvement of our health. Most are used as salads or eaten raw but sometimes they are cooked or even fried and added to other vegetables. Flowers in most cuisines are used in salads, cocktails and desserts to bring out its ornamental value and it is edible. It should be noted that not all beautiful flowers are edible some are potent and could be poisonous.

## 3.7 WILD EDIBLE SEEDS.

Generally, seeds are known as the part of the plant that is grown to yield another plant of the same species. This is not entirely true. Seeds can also be ingested. An edible seed is one that can be ingested without causing any harm. The most recognized edible seeds are cereals then legumes nut and spices. Seeds that are regarded as edible nuts have been discussed above. They include almond, Brazil nuts, cashew nut, Acorn, chestnuts, hazelnuts and macadamia (Dan Dourson & Judy Dourson, 2019).

Examples of wild edible seed include:

- Amaranthus seeds- these seeds contain a substantial amount of protein, fiber and macronutrients including, manganese, magnesium, iron and phosphorous. They are considered very healthy as they are gluten- free. In this case, the cholesterol levels are reduced and there are high chances of reducing weight. Even though the popularity of these seeds was just gained recently, it was a very popular grain during ancient times. It went to an extent whereby it was a staple food among Inca, Maya and Aztec groups. This fruit does not however technically a real grain though it contains the same set of nutrients as other grains and can be consumed in the same way as other cereals. It is consumed with a variety of dishes as it has a nutty, earthy flavor. The seeds also contain antioxidants that protect the body from harmful free radicals that can damage crucial cells and contribute to the occurrence of chronic diseases. The antioxidants that are in amaranth are known as phenolic acids. The ingestion of Amaranth can help reduce inflammation in the body. Amaranth is soaked and left for one to three days to sprout then cooed. This is done so as to make the seed easier to digest. They can be added to smoothies, mixed with soup, mixing with fruits and can be used instead of rice. *Amranthus retroflexus* commonly known as pigweed is a plant species in the family of amaranth. Its seeds are not difficult to harvest. One can decide to ingest raw or cooked versions of this seed. They can also be roasted or even popped like popcorn. Just like the rest of the amaranth family member plants, *Amaranthus retroflexus* seeds are of high nutritional value.

- Prickly pear seeds- as seen earlier in this document, prickly pear fruit are cacti fruits that are edible. Once peeled, one eats the fruit. The fruit contains the seed that is very hard to break through but one can decide to swallow as is or throw it out. Researchers have concluded that these seeds are rich in nutritional value. The contents of the seeds are fat, ash, protein and fiber. The seeds contain also a substantial amount of minerals such as iron, magnesium, phosphorous, zinc and potassium . All these minerals are very important to the human body.
- *Borago officinalis* commonly known as borage seeds- these seeds are considered not only edible but also herbal. The seed can be pressed to produce oil that contains omega-6 fatty acids and afterward sold in the market as a GLA supplement. It contains various nutrients, such as B vitamins, beta-carotene, fiber and some amounts of trace minerals.
- *Brassica juncea* Seeds commonly known as Mustard green seeds- the seeds are very common spices. It is a major ingredient in a beverage known as prepared mustard in which the seeds are ground and mixed with vinegar and other liquids. They can also be pressed to produce mustard oil.
- *Chenopodium album* seeds commonly known as Lambs quarters seeds- are commonly found close to rivers and streams. This is another type of pigweed. Some lamb quarters seeds are used as a food staple. The seeds are considered very nutritious and one can use in recipes. Their seeds can be milled to produce flour that can be used to bake bread. The sprouts of lamb quarters seeds can be used as additives to a variety of diets. Though small, a substantial amount of these seeds are a potential source of energy that can sustain a person. They are nutritious as they contain a substantial amount of vitamins and minerals which include iron, phosphorous, folate and magnesium and also vitamins A and C.

- *Salvia hispanica* commonly known as Chia seeds- the seeds of the plant are edible. It is native to the individuals in central and southern Mexico. Chia seeds can be used when they are ground or whole for nutritious foods and beverages. Chia seeds have nutritional value. Once dried, they contain water, carbohydrates, protein and fat. They also contain a substantial amount of minerals ranging from magnesium, calcium, iron, manganese, zinc and phosphorous. They also contain huge amounts of Omega-3 fatty acids that are very important for brain development and fiber that help in digestion. Chia seeds can be processed to produce oil that can be used for cooking. In other cases, individuals may decide to sprinkle chia seeds on top of other foods. One can also mix chia seeds with smoothies, breakfast cereals, yogurt, tortillas, granola bars and bread. One can also decide to soak these seeds in water and consume them as they are or they can also decide to mix with milk or even juice. It is a major ingredient used in making chia seed pudding that is similar to tapioca pudding along with a sweetener and a certain type of milk.one can also decide to eat chia seeds as they are.
- Pickerelweed seeds- this plant grows in places with shallow freshwater. Its seeds can be eaten raw and even tastier when roasted. The seeds can be tossed into dough used in making bread. They can also be ground and the flour used in baking.
- Wild sunflower seeds- the seeds of the wild sunflower are much smaller than those that are commercially produced. They are consumed raw. Fats are a major constituent of these seeds. They can, therefore, be ground and the flour mixed with another flour and used for the purpose of baking.
- Garlic mustard seeds- they go well with spicy food. One can toss the seeds on a meal before consumption which makes the spicy food tasty. An individual can also eat the seeds raw.
- Wild black cherry- even though the seeds are edible, they should be consumed in small proportions as the seeds contain a compound that can be used to make cyanide.
- Peppergrass seeds- can be used instead of black pepper. The Chinese people use these seeds in the preparation of herbal medicine.

- Blue vervein seeds- the seeds of the blue vervein plant are best eaten roasted. They can also be ground to produce flour. They portray a bitter taste.
- Crimson clover seeds- it grows in the wild, especially in meadows and lawns. The seeds can be eaten raw. One can decide to allow the seeds to sprout and use them in salads and sandwiches.
- *Zizana aquatic* commonly known as wild rice- scientists do not consider it as rice but rather seeds. One can eat the seeds as is or can decide to cook. The Native American communities rely on this rice as they treat it as their source of food. The seeds can also be ground to produce flour. The flour is not used for baking but rather mixed with other flours or can be used when one wants to thicken the soup.

This shows that one should not disregard the seeds of certain plants. They provide substantial amounts of nutrients that are helpful to the human body. Some act as spices while others as flour and others are food and beverage additives.

## 3.8 WILD LEAF VEGETABLES

They are commonly referred to as wild greens and they were gathered while they are still young and soft because when they mature they become hard and have a bitter taste. There are many leaf vegetables that grow in the garden or farm as weeds. Most people remove them from the farm as they see them as a nuisance, instead, they should cut them and feed them to the livestock if they don't eat them themselves. Not all weeds are leaf vegetables as some could be harmful to the body as the mint contain toxins (Tardío, J., Pardo-de-Santayana, M., & Morales, R., 2006).

*Lamb's quarters:* Scientifically known as *Chenopodium album* is a common weed that grows in most home gardens as it loves land that has been disturbed. It is an opportunistic plant. It is a weed that are found in Japan, South Africa, Europe, India and in America. They are also referred to as wild spinach. It contains a lot of vitamins and minerals. The leaves of the lamb's quarters are alternate and triangular. The margins are coarse and shallow- lobed. They have a whitish- gray coating and is powdery is visible in young leaves. The coating is usually waterproof. They grow between one to one and a half meters in height. It contains a red pigment as grows. While harvesting them the top nodes are easily pinched by hand. The immature tips are tender enough and only rinsing is needed. The lower part of the stem is fibrous and only leaves are picked. When the leaves are picked they regrow fast and younger leaves are produced. They can be eaten raw. They can also be steamed in lasagna. They contain a lot of oxalates but a considerable amount of minerals and are not harmful to the body. The oxalic acid may be harmful to the body hence it taken in small amounts. They contain iron and potassium. They can also be cooked in soups and stews.

**Figure 3.28:** Lamb quarters. top nodes can be pinched and consumed raw after rinsing.

*Source: https://plants.ces.ncsu.edu/plants/chenopodium-album/*

*Mallow:* Scientifically known as *Malva sylvestris*. It is also known as malva which is derived from the Greek word "malakos" which simply means "to soften". It is native to Europe, western Asia and the northern region of Africa. They grow well in fertile soils that are loamy but do not prefer acidic soils. They grow up to 1.5 meters high. The leaves are lobed around the stem and alternate along the stem length, the leaves are five- lobed. The stem is hardy and fibrous. They contain vitamins A, B, C, E, essential fatty acids, fiber, calcium, magnesium, selenium, zinc and flavonoids. The young leaves are harvested and can be consumed raw or sometimes used to make smoothies, in salads only very young leaves are used. They can also be steamed in vegetable dishes. In some groups of people, they are deep- fried.

**Figure 3.29:** Mallow plant. grow well in fertile soils that are loamy.

*Source: https://www.gardeningknowhow.com/ornamental/flowers/common-mallow/controlling-mallow-weeds.htm*

*Dandelion:* Scientifically known as *Taraxacum officinale.* It is smooth-margined or toothed. They grow in clay-loam or loamy soils and prefer little sunlight. They mostly grow in early spring when there is enough warmth and sunlight. They can grow on lawns, kitchen gardens vacant lots or sometimes along roadsides. The young leaves of dandelion are harvested before the plant flowers. The leaves are tender and have a slight bitterness. After it flowers leaves become tougher and bitterness increases. The young leaves are added to salads. When being cooked it is first boiled and the first cooking water is poured out to get rid of the bitterness, it can also be used in making of soup. The flavor of dandelion is both earthy and bitter which is almost similar to endive. Dandelion leaves are a great source of vitamins A, C, K and E. they also contain folates and trace amounts of B vitamins. It is also a great source of minerals such as iron, calcium, potassium and magnesium. It also contains a lot of antioxidants such as beta-carotene and polyphenols. They have many benefits to the body; help lose bodyweight, kills cancer cells, boost the immune system, support the secretion of bile and the fibers in the leaves help prevent constipation.

*Nettles:* The most common leaf vegetable among the nettles is the stinging nettle. It is scientifically known as *Urtica dioica* it comes from the Urticaceae family. It is commonly referred to as; nettle leaf, nettle or stinger. It is native to Europe, northern regions of Africa and western Asia. It has now spread worldwide including the Americas and New Zealand. It is known for its stinging leaves that are very irritating. It can grow to heights of up to two meters high. It spreads vegetatively with yellow rhizomes and grows in large clusters. They contain numerous stinging and non-stinging hairs. When the hairs come in contact with the skin it becomes irritating and it gives a burning sensation that is very painful. The leaves cannot be eaten raw as the stinging hairs have to be removed before consumption hence, they cannot be used as salads also. The plant is processed by steaming, drying, macerating or sometimes prepared in veneer or alcohol solution. Their common habitats include forests, foothills and coastal regions. When harvesting nettles a person needs to protect themselves by using tongs or wearing protective gloves. They are dried and it is used to prepare tea. They can be cooked in soups or in some African dishes it is prepared as a vegetable stew. It contains beneficial minerals such as calcium, magnesium, iron as well as potassium. It is also rich in vitamins A and C. They contain a lot of protein and helps in liver enrichment. They are known to grow vigorously and need to be controlled, they are pruned and the root runners dugout to prevent them from covering large areas of the field or garden.

**Figure 3.30:** The stinging nettle. contains hairs which when they touch the skin they cause a burning rash.

*Source: http://ravensongherbals.com/herbs/nettle-stinging*

*Amaranth:* The most common of the amaranths is the *Amaranthus hybridus.* They are commonly referred to as pigweeds. They mostly grow in the summer and autumn seasons. It produces a lot of seeds and is regarded as invasive as it grows in large numbers. That can be both an advantage and a disadvantage. The seeds are also edible and when they grow they grow as weeds and they absorb nutrients from the main plant in the field. Their taste is similar to spinach but it is much stronger. It is used to prepare vegetable stews, used also in salads and it can also be stir-fried. It contains iron, calcium and potassium minerals.

*Mustard:* The wild mustard scientifically known as *Sinapis arvensis L.* its leaves are smooth and kidney shaped. They have hairs. The lower leaves are irregularly lobed and toothed. Upper leaves are alternate with saw-tooth margins and their tips are pointed. It is found in many parts of the world especially on the seasides and on the coasts. It is hot and spicy. The young leaves are mostly eaten raw and sometimes eaten as salads. As they get older the leaves tend to be hairier and they are best eaten when cooked or steamed. It is used to make stews and sometimes used for making soup. They contain a lot of minerals including calcium, potassium, manganese, zinc magnesium and selenium. They also contain little amounts of protein and trace amounts of fat. They contain vitamins A, C, K and E, and also folic acid. They also contain good amounts of fiber and they are good for preventing constipation. They are also good sources of antioxidants.

**Figure 3.31:** Dock plant. have long narrow and pointed leaves.

*Source: https://www.healthbenefitstimes.com/dock/*

*Dock:* Scientifically, known as *Rumex crispus.* It is also commonly known as curly dock, yellow dock or curled dock. The leaves are alternately arranged, long and narrow whereby the tips are pointed or sometimes round with margins being wavy. They tolerate almost all types of soil. They can withstand drought and they are known to have seeds that can remain viable in the environment for up to 50 years. The leaves are edible while still young where they can be eaten in salads. The leaves when cooked are first boiled in water then the first water is poured, they are then rinsed using running water, they are used to make stews or added to other vegetables. They are rich in vitamins A and C. They also contain tannins and oxalic acid.

*Chickweed:* It is scientifically known as *Stellaria media.* It creeps along the ground and is evergreen. They can also be found during winter below the snow. It is known as the chickweed because chicken and ducks love it. It is native to Eurasia but is now common throughout the globe. Its leaves are oppositely arranged and oval in shape. The leaves are very succulent. The leaves contain a lot of chlorophyll and that makes them have an effect at the purification of blood. It can be eaten in salads or it can be dried and used to make tea. They can also be eaten raw, used also for making juice and can also be steamed. Dried chickweed lose potency very fast making them lose them not unsuitable. It is a good source of vitamin C.

*Purslane:* It belongs to the family Portulacaceae. It is scientifically known as *Portulaca oluracea.* Its height is only 40 cm from the ground. It is a succulent plant that grows in many parts of the world. It is mostly found in the East Mediterranean countries. They grow in many habitats including salty and soils that lack nutrients. It contains vitamins A and C. It also has minerals i.e. magnesium, manganese, zinc, potassium, It also contains trace amounts of vitamin B1, B2 and B3. It also has trace amounts of folate, copper and phosphorus minerals. It is also said to contain omega-3 fatty acids. The whole plant is edible before it starts flowering. The young shoots are used in soup making. It has a smooth taste and sometimes sour or salty. It can be eaten raw or cooked and eaten as a stew.

*Chicory:* It is scientifically known as *Cichorium intibus.* It grows along roadsides but is mostly found in wild fields. It is a drought- resistant weed and prefers sunny areas. The leaves are tooth- shaped but is tall growing. The young leaves can be used in salads while when they become mature they are cooked or steamed. They can be dried and used in tea.

*Fennel:* Its scientific name is *Foeniculum vulgare.* It grows mostly in the Mediterranean region and grows in the coasts, foothills and riverbanks. The young shoots are edible and they can be eaten raw or sometimes added to salads. They are also steamed or cooked and integrated into other dishes. It contains minerals such as calcium, iron, magnesium and potassium. It also contains vitamins A, C and B-6. It also contains traces of copper, selenium, niacin, pantothenic acid, folate, choline, lutein, vitamin E and K.

*Grape leaves:* Known as *Vitis species.* It grows where wine grapes are found.it is found growing as it weaves itself onto other trees and bushes. They are mostly used for food fillings. They can also be canned, steamed or fermented. Young leaves are harvested early and taken raw. It can also be cooked together with other vegetables. As the plant matures it becomes less palatable. It contains a lot of fiber and is a good source of vitamins C and K, beta-carotene and calcium.

**Figure 3.32:** Grape leaves. It grows by weaving itself along other trees.

*Source: https://www.thankyourbody.com/recipe-wildly-fermented-pickled-grape-leaves/*

*Milk thistle:* Scientifically, known as *Silybum marianum.* It is commonly known as Mary thistle, milk thistle or holy thistle. It is native to the Mediterranean region (southern Europe and northern Africa), Asia Minor and southern Russia. The leaves are oblong with spiny edges. They do not have hairs are shiny green and their veins are milk-white. The leaves when being harvested have to be trimmed around the exterior to remove the thorns. They are eaten steamed or cooked. It can also be chopped and added to salads but not most desirable. It contains anti-oxidants.

*Miner's lettuce:* It is known as miners because it was mostly used by California gold rush minors. The leaf is round and hardy. It is mostly found along the foothills, woodlands and coasts of California. It contains high amounts of vitamin C which help in preventing scurvy. It also contains omega-3 fatty acids.

*Plantain:* Found in the genus Plantago. It has a long slender oval -shaped leaf. Young shoots are soft and appetizing when eaten as a raw salad. It can also be cooked. The older leaves are bitter. Young leaves contain beta-carotene, calcium and vitamin C.

*Salad burnet:* It is scientifically known as *Sanguisorba minor.* It grows in almost all climatic zones and is drought tolerant. It grows throughout the year. It enjoys soils that re rich in limestone. They are quite short growing about 40-100cm in height. The leaves which look like cilantro grow on both sides of the stem. The leaves are harvested by pinching then sliding down the stem. It can be blended into a salad dressing, added to a salad or sometimes eaten raw.

*Violet leaves:* Scientifically called *Viola odorata.* They have spice and or nutty flavor. They have a scalloped heart- leaf. They grow like weeds in gardens or lawns. They are also found close to water sources. The young leaves can be eaten raw. They become bitter as they mature they can also be dried and used to make tea.

*Watercress:* *Nasturtium officinale* is its scientific name. It grows in aquatic areas i.e. ponds, slow- moving water and streams. They are a good source of isothiocyanates that are responsible for their pungent flavor. The leaf has a spicy and peppery taste. They can be eaten raw or can be used to make watercress soup.

*Strawberry blight:* Scientifically, it's known as *Cenopodium capitum.* It grows mostly in Europe and be easily seen in monasteries. Its appearance is almost similar to lamb's quarters although its leaves have a red tinge at the bottom and are frosted on top. The leaves are harvested while still young and tender. They can be eaten raw or added to salads.

*Sheep sorrel:* Also commonly known as sour dock. Scientifically known as *Rumex acetosella.* They are weeds that grow on cultivated land or lands with human activity. It spreads on the ground with many creeping roots. Leaves have an arrow shape while stems have a reddish color. When eaten raw, the leaves have a sour lemony taste even when mixed into a salad. They are sometimes mixed with other greens and eaten as vegetable stews. They are best eaten while still young. They can also be chopped, added to green onions and cooked with mashed potatoes.

*Taro:* It is sometimes called kalo. Its scientific name is *Colocasia esculenta.* It has heart shaped leaves. It grows in tropical and subtropical areas. It is commonly known for its root which contains a lot of starch. The leaves of the taro are never eaten raw as they are poisonous. They are then cooked before being consumed. They are known to contain minerals such as potassium, calcium, iron, magnesium and phosphorus they also contain vitamins A and C and folates. They also contain proteins, fats and fiber. They are also known to contain antioxidants which help in disease prevention. They contain dietary nitrates.

**Figure 3.33:** Taro. It is cooked before being consumed. poisonous when eaten raw.

*Source: https://www.thecoconet.tv/cocoblog/health-benefits-of-taro/*

*Wild growing ferns:* There are three ferns that are known to be edible. These include; the ostrich fern (*Matteucia struthiopteris*), the lady fern (*Athyrium filix-femina*), and bracken fern (*Pteridium aquilinum*). They are mostly found in rich hardwood forests. They grow from between 1-2 meters in height. While harvesting them for consumption the fiddleheads but not the full -grown fond are harvested. The fiddleheads are smooth and lack scales. They are collected in the middle of spring or in late spring. Harvest while they are still tender when the fronds are yet to unfurl. The fiddleheads are mostly canned and used later. They can be eaten raw as they are crisp and sweet. They can also be added to salads. They are a good source of vitamin

B2, phosphorus, potassium and copper. It also contains good amounts of vitamin B3 and C and manganese.

## 3.9 WILD EDIBLE ROOTS

Roots are part of plants that lie beneath or on the surface of the ground. Roots that appear above the ground are said to be aerial or aerating for those above the water surface. Roots do not contain leaves nor nodes on its body. Plant roots form the basic foundation of a particular plant. Plants obtain nutrients especially, from the soil. This explains why roots are in fact the most important part of the plant. Without roots, plants seize to exist. Some plants would however prefer to use roots for the storage of food and water. Roots are located under the soil and hence have low risks of losing water to the ecosystem. On the other hand, roots can be used for the storage of plant food. Since they come into contact with important nutrients some plants may prefer storing their food in the same position. Such plants include arrowroots and yams (Merritt Lyndon Fernald & Alfred Charles Kinsey, 2012).

These roots also feed the human population. The same nutrients that they provide for plants may be beneficial to humans. Some plants that thrive in the wild have very nutritional roots. Though many countries have made it illegal to dig up forest vegetation, there are plant species that would provide food to ethnic and cultural communities. In the past, people had the will to protect forests and their wild growth. This made it possible for man and plant to thrive in co-existence.

Wild edible roots are known to thrive mostly on riparian land and river banks. Plants such as the arrowroots take after in inhabiting region s that experience high water levels. In the deserts, plants that bear produce on roots are most important in the provision of water to communities living in the desert regions.

The following are some examples of wild yet edible roots.

*Daylilies:* This is a beautiful flowering plant of the genus *hemerocallis.* The plant is however not a lily as the name depicts. Daylilies is a name that describes how the plant would thrive flowers that would last only for a single day. The flower would be bourne out of the plant in the morning and would wither by the following nightfall. The plant is native to Asia in countries such as China, North Korea and Japan. They have a fibrous root system which allows them to tap water and nutrients from close to the surface. Daylilies use their roots to store water and food. The roots are used by humans as food and for medicinal purposes. The roots are mostly trimmed off and sun- dried before being boiled for traditional medicine. Older daylilies root are inedible and even have a poor taste (Merritt Lyndon Fernald & Alfred Charles Kinsey, 2012).

*Cattails:* Cattails grow on riparian land or river banks. They inhabit fresh and slightly brackish water hence said to be aquatic or semi- aquatic. Their leaves are quite long and tapering. The plant is dispersed by wind. Due to its wet conditions, the plant is harvested by removing the mud around the plant and slowly pulling out the roots. The roots are then washed off with water, removing the mud making it quite tasty and sweet. The root can be ingested raw. Other cuisines suggest that it is either cooked or boiled.

*Chicory:* This is a perennial and herbaceous plant that belongs to the dandelion family. It is scientifically known as *Cichorium intybus.* It bears bright blue flowers that are rarely white or pink in color. The leaves of chicory are unlobed. They are of the lanceolate structure and ever remain stalked. Root extracted from the plant is used as a substitute for coffee. The roots undergo various processes to be suitable for human consumption. They can be roasted, cooked or grounded to be used as an additive. It is used in the preparation of a beverage delicacy in India called the Indian filter coffee. In France, it is used in the preparation of ricore. Chicory is also used in the beer industry whereby it is used for brew making. The chicory roots are roasted and used to add flavor to stouts. Moreover, chicory roots can be cooked and eaten directly.

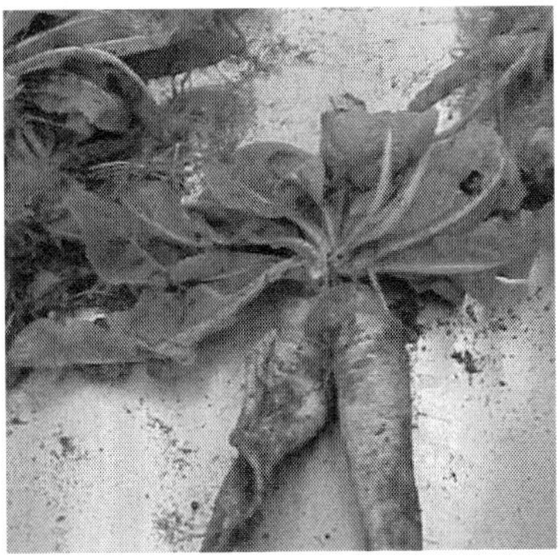

**Figure 3.34:** A chicory root; it is used as a substitute or rather an additive to coffee.

*Source: https://www.researchgate.net/figure/Chicory-plants-Cichorium-intybus-L-showing-outer-leaves-and-tuberous-roots_fig1_237548269*

# 4

# *Edible Aquatic Plants*

## CONTENTS

# 4.1 INTRODUCTION

Ponds, wetlands, swamps, oceans and rivers all around the world are home to various edible plants. Some freshwater edible vegetation are wild weeds, while others help control the fragile ecosystems of rivers or streams by providing food as well as nutrients for people, fish, and wildlife. The freshwater edible plants also offer shelter for marine fauna and terrestrial animals. Most of these plants are completely or partially immersed in water. A number of aquatic vegetation is edible and offers the same valued nutrients available in leafy green veggies such as spinach. These plants may usually be bought fresh on the farmers market or can be grown in the backyard so long as the circumstances are right.

Thousands of marine and semi-aquatic plant species live in various habitats of rivers, including fast-flowing streams, swamps, marshes, and slowly moving acidic black water rivers. For large rivers like the Amazon, occasionally key waterways flood and cover massive areas of the surrounding forests, greatly changing the landscape and forming new temporary habitats where edible plants can thrive and be consumed by local residents.

Nowadays, edible aquatic plants can be grown at home for consumption or lawn beautification purposes. You can either place them in airtight containers filled with water or if you have a small pond in your compound the better. Additionally, with advancements in technology some scientists are using these marine plants for novel functions like water purification, commonly known as floating-type edibles aquatic plant purification systems (FEAPPS).

# 4.2 POPULAR EDIBLE AQUATIC PLANTS

**Figure 4.1:** Wasabi is an edible aquatic plant that originates from Japan.

*Source:* https://www.quora.com/What-are-the-plants-that-grow-in-water-and-we-eat

### 4.2.1 Wild Rice

There are four kinds of wild rice in the Zizania genus (*Zizania aquatica*; *Zizania palustris*; *Zizania latifolia* and *Zizania texana*), which is a set of edible grasses found in small lakes and shallow water with a slow flow. Usually, only the flowering top of rice is higher than the water's surface. Northern wild rice (*Zizania palustris*) is native to North America, an annual crop native to the northern forests of Great Lakes region in Alberta and Saskatchewan (*Z.aquatica*)-also annual, grows around the St. Lawrence River and the Atlantic as well as Gulf Coasts.

At the same time, Texas wild rice or (*Z. texana*) is a perennial crop found only in a small part of the San Marcos River of central Texas, while Manchurian wild rice or (*Z. latifolia*) is a native perennial from China. The annual *Zizania palustris* seeds are the most widely harvested grain. Rice is rich in protein, amino acid lysine and dietary fiber, but low in fat. Same as true rice, wild rice is gluten-free. It's also a rich source of minerals phosphorus and potassium, including vitamins riboflavin, niacin and thiamine.

### 4.2.2 Watercress

Watercress is a quick-growing aquatic and semi-aquatic perennial plant that grows from Europe to Central Asia. It's also among the oldest leafy vegetables eaten by humans. These crops are members of the Cruciferae or Cabbage family and are biologically related to cress and mustard, all of which have a peppery and rich flavor. The hollow stem of watercress floats and the leaves are pinnately compound. Watercress grows white and green small flowers. The cultivation of watercress is feasible both on a large scale and on a garden scale. Watercress is a semi-aquatic plant, very suitable for hydroponic cultivation, the most dynamic environment for them being in slightly alkaline water. If not harvested, watercress can reach heights of 50 to 120 cm. Commonly sold as sprouts, their edible shoots get harvested a few days after germination.

Like many other crops in this family, when the crops begin to bloom, the leaves of the watercress become bitter. Apart from vitamins A and C, watercress also contains a lot of folic acid, iron and calcium. In some areas, watercress is considered a weed, and in other areas, it is considered an aquatic herb or vegetable. Eating watercress has numerous benefits, such as acting as a moderate stimulant, a source of antioxidants and phytochemicals, diuretics, expectorants and digestive aids. It also seems to have cancer-suppressing effects and is widely considered to help prevent lung cancer.

### 4.2.3 Water Spinach

Water spinach, commonly referred to as Ipomoea aquatic, is a semi-marine tropical plant that grows in the form of leafy vegetables. Due to widespread cultivation, its precise wild distribution is unknown, and the species is common throughout tropical and subtropical parts of the world. Some of its common names include water spinach, spinach, swamp cabbage, kangkung, kangkong, and water morning glory. Aquatic spinach grows in water or wet soils. Their stems are 2 to 3 m or longer and hollow, allowing the plants to float and establish themselves at the nodes. The leaves vary in shape from sagittate (typical) to lanceolate, measuring 5 to 15 cm long and 2 to 8 cm wide. Their flowers are trumpet-shaped and 3 to 5 cm in diameter, usually white.

It's most typically found in Southeast East and Asia. Since it grows wildly in waterways without requiring too much care, the plant is also commonly used in Chinese and Malay cuisine, particularly in rural regions. It is frequently mistaken for watercress that grows under similar conditions, but they are quite different plants.

Water Spinach has further been introduced to the United States, and its high growth rate makes it an environmental issue, particularly in Texas and Florida. Though it has been designated as a "harmful weed" by the United States Department of Agriculture, this plant is harmless when eaten and is comparable to traditional spinach in nutrition. The vegetable is also commonly used as an ingredient for Southeast Asian meals, where the leaves are typically stir -fried with garlic, ginger, chili pepper, and dried shrimp paste among other spices.

### 4.2.4 Wasabi

Wasabi is a member of the Brassicaceae family, which includes cabbage, horseradish and mustard. Known as "Japanese horseradish," its root is used as a spice and has an extremely strong flavor. Its hotness is more like that of a hot mustard than the capsaicin in chili pepper, producing vapors that irritate the nasal passages more than the tongue. The plant grows wildly along stream beds in mountain river valleys in Japan. There are also other species used, such as *W. koreana*, and *W. tetsuigi*. The two main cultivars in the marketplace are *W. japonica* cv. 'Daruma' and cv. 'Mazuma'.Wasabi is generally sold either as a root (real wasabi), which must be very finely grated before use or as a ready-to-use paste. Fresh leaves of wasabi can also be eaten and have some of the hot flavors of wasabi roots.

You can eat them as a mustard salad by marinating them with salt and vinegar-based seasoning overnight or boiling them quickly with soy sauce. Their leaves can further be beaten into a paste and then fried into thin slices. Wasabi is usually eaten with sashimi or sushi, commonly served with soy sauce. The latest research has revealed that mustard contains wild chemicals that can be utilized against some cancer cells. This special root can further be applied for oral health and infections. It's been recommended that Wasabi can help inhibit cardiovascular diseases such as stroke, high blood pressure and heart disease. There are various other health benefits of Wasabi roots such as preventing asthma, arthritis, diarrhea, allergies and osteoporosis.

## 4.3 GROWING EDIBLE WATER PLANTS AT HOME

The water garden may provide interesting plants and beautiful flowers or leaves for the beauty of consumption. For small space gardeners planting vegetables, having a water garden for edible plants is a low maintenance preference. There are many plant options suitable for the water garden. You can consume the leaves of water ferns, chameleon plants or water spinach. Its leaves, seeds and tuber roots are the main pillars of Asian cuisine. Simply add water celery leaves to the soup to get a mild celery taste, or use water mint leaves like other traditional mints. You can grow water plants in any shallow container which can hold water. But if you would wish to place them inside a pond, consider using a container with drain holes on the sides or bottom, this is especially applicable for water chestnuts (Angier, 2001).

Adding edible pond plants is an interesting way to make your backyard pond more interesting. You can add green sprigs of aquatic mint grown from home to your iced-tea for increased flavor, or crisp watercress to the salad. These fresh greens from the garden increase the outdoor garden appeal and encourage a healthy lifestyle. Many people do not know that there are various plants from the water garden habitat that have already made their way to the table. We're not just talking about salads or bean sprouts, but also nutritious snacks, such as fresh green lotus seeds, which can be consumed on the go.

The demand for favorable environmental solutions is higher than ever. Customers are gathering rainwater, utilizing solar panels, and using swamp filtration systems for the backyard ponds as well as water features. In addition to choosing a healthy environment, customers also choose a healthy lifestyle by adding fresh products to the table. Presently, there is a great demand for palatable pond plants.

Adding edible aquatic plants to your backyard will indeed improve the home's ambiance, apart from promoting good health. Although most plants flourish in traditional gardens, there are multiple other plants which flourish in water gardens, swamps and rainforest environments. Taro and rice are among the first aquatic plants to be cultivated many years ago. They are ideal for making fresh salads and soups, stews, curries, snacks, and stir-fried dishes (Angier, 2001).

### 4.3.1 General Planting Tips

When building up a pond or some water feature that covers space for cultivating edible plants, preferably choose a sunny spot without tall trees nearby. Furthermore, make sure to pick an easily accessible area and plant edible crops in the shallow water of the pond, chosen plant pockets and swamp regions. When planting palatable pond plants, use fertile loam or Turface MVP for the pond and allow green leafy vegetables to grow just above the waterline. Furthermore, consider using edible crops in swamp filters, for this, water celery and Taro make excellent choices. To ensure that the clogged filter is not jammed, always dig out the clogged filter 2 feet deep and apply a mixture of Turface MVP at 50-50 ratio, including pea gravel from an eighth- inch to quarter inch.

Using pea crushed stones bigger than 1/4 inch may impede plant growth, or crush crop roots. You may plant edible pond crops in fabric flower pots, and then add them in the river bed of the pond to recycle the water back into the pond. It's the ideal way to bring healthy oxygenated water into the plant while cleaning the water and ensures it remains fresh. You can include some fish to the pond where possible to assist in wild fertilization while keeping mosquito larvae away. Fertilizers can be also added, but don't use fish medicine or chemicals in the pond, otherwise, the plants that have been eaten will become inedible.

### 4.3.2 Tips for Growing Different Edible Water Plants

*Chestnuts*: First, the best period to grow water chestnuts is during springtime. Place 2 to 3 inches of soil in an 18-inch basin that's at least 6-inches deep. Separate the bulbs by 1 to 2 inches. Cover the container with 1 to 2 inches of soil. Add the soil gently. For pond cultivation, cover with the layer of pea particulate. If you do not have any outdoor space for the pond, consider adding a nice small water garden as the focal spot on the deck or terrace. Next, pour water on the plant to fill up the container.

Always ensure the water level is roughly 1-inch above the soil (In a swamp or pond, immerse the container in the water so that there is 4 to 6-inches of water above the edge) Put the plant in a place full of sunlight. Fertilize the area with some manure every 3 weeks. These techniques ensure that the aquatic garden plants in the lawn remain healthy and beautiful. Ultimately, harvest the matured water chestnuts in around 6 months. Once the leaves turn brown, remove the plant from the soil. Wash away the soil and wipe the leaves from the chestnuts.

*Sacred Lotus:* This is a perennial herb with a length of 1 to 1.5 m. it features large round leaves that are formed on long stems. The flowers are incredibly beautiful and usually white or pink. In dams found in warm climates, the sacred lotus is most likely to be intrusive. Most of the lotus are edible; its leaves can be used to cover steamed food, or consumed as green vegetables when young, the roots of the tubers are similarly used for cooking, the petals are floated within the soup or spread in the salad, and then baked as roasted seeds.

In order to meet the germination demand, the lotus needs fertile soil, a long growing season and sufficient sunlight in a warm environment. Upon establishment, they are quite hardy but need constant care and warm growth conditions to sprout. Choose a broad shallow pot for germination, a container measuring 33 cm x 23 cm is ideal. For bigger ponds, old synthetic clothes, baskets may be utilized.

Mix the heavy and fertile garden soil together with the potting mix on a 50:50 ratio, then proceed to fill up the pot to an average depth of 10 cm. The tubers must be in a partially dormant state upon arrival. Handle the tubers carefully so as not to break the germination tip, also known as the "eye". This is the section of the tuber whereby leaves shall sprout. The "eyes" should not be concealed by gravel or soil. Add the tubers horizontally on the earth and spread the "eyes" to swell from the earth and gravel. Conceal the remaining part of the crop with soil. Following the sowing, include a layer of gravel atop the cover to ensure the water remains clean and inhibit the soil and water plants from floating mutually. A few flat rocks could be needed to decrease buoyancy. Add the pot beneath the water level until the plant sprouts, since the nearer the pot is next to the water, the greater the temperature. Progressively, lower it over throughout the coming few weeks, until the pot is covered with water between 10 and 23 cm.

*Hardy Water Lilies*: In addition to garden ponds, beautiful water lilies can grow in bathtubs on the terrace or in ponds that come in various sizes. They develop from tubers grown in underwater pots, with round leaves and star-shaped flowers floating on the surface. These flowers bloom during the morning, around sunset, lasting for 3 to 4 days before sinking into the earth. Flowers show up from spring season to autumn. These plants hibernate in winter and might remain in the water or get stored.

For growing requirements, at an average water depth of roughly 30 to 45 cm, planting in sufficient sunlight is very important. Use a wide and shallow pot or container. A suitable size is 30cm x 45cm wide x 15cm deep. In case there are some drainage holes present, preferably line the plant pot with burlap in order to keep the soil in the dish. The rhizome matures horizontally throughout the container, so it should start from one side of the vessel. The rhizome is grown at an angle of around 45 degrees, the tree crown is exposed, often pointing toward the middle of the vessel. After planting, spread a layer of gravel. You can first reduce the plant to a depth of 15 cm, and then as the plant grows to a depth of 30 to 45 cm. Planted flower pots should be lowered into the pond at an angle to allow air to escape. Regular fertilization throughout the growing season will ensure flowering.

*Pickerel Rush:* This is a perennial pond plant featuring smooth green leaves and charming blue flowers that grow in summer. Pickerel rush is quite hardy and simple to grow, as long as it doesn't dry out. Hyacinth-like flowers will continue to bloom throughout the summer. The plant provides adequate nectar for bees or butterflies and habitat for water fauna. It's a striking addition to water gardens. The nutrient-rich starchy seeds may be eaten directly from the plant, or they can be dried and included in cereals, which tastes more like swelled brown rice.

Seeds may also be stored dry, then boiled or baked to improve flavor or grounded into flour. Young leaf stems may be cooked green or consumed raw as a salad. This emergent aquatic plant matures in shallow waters or swamp plants. Simply place the plant in a shallow, sunny container with a water depth not exceeding 20 cm. Fill the container up to 10 cm using thick fertile garden soil and potting mixture at a ratio of 50:50. After planting, spread a coating of gravel cover above it. Frequently, divide the plants to prevent them from becoming too large.

# 4.4 HOW EDIBLE AQUATIC PLANTS CAN PURIFY WATER

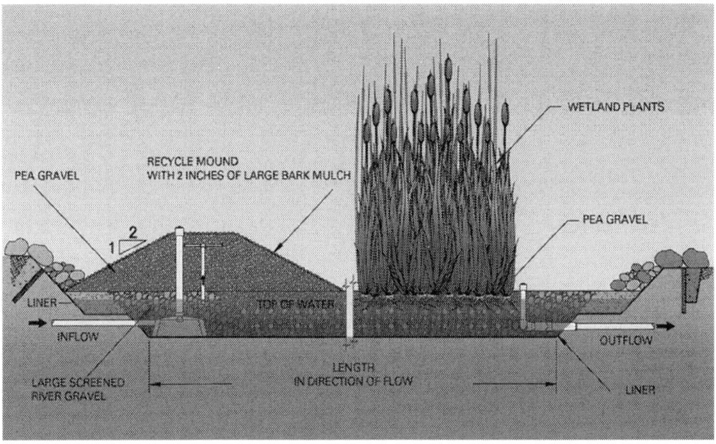


**Figure 4.2:** Aquatic plants can wildly remove toxic chemicals from water.

*Source:* http://butane.chem.uiuc.edu/pshapley/Environmental/L32/3.html

The water purification features and attraction impact of aquatic fauna in the floating edible aquatic plant purification system (FEAPPS) have been demonstrated. FEAPPS is a defined ecological engineering technology in which aquatic plants are floated on the water, while their roots are suspended beneath the water surface. At each eutrophic level, the concentration levels of SS, as well as turbidity in FEAPPS, is lower compared to the control system, since the presence of aquatic edible plants facilitates root filtration, thereby removing soil sampling (SS) and turbidity in FEAPPS (Department of the Army, 2009).

The average ratio of crustacean biomass as well as transparency in FEAPPS is much higher compared to the control system, besides the correlation between these values is also quite high. It shows that the FEAPPS application helps to preserve a greater feed of crustaceans being filters, apart from achieving higher transparency within the system. The environmental restoration technology using ecological engineering can be an energy-saving, affordable and sustainable purification technique, so it has recently attracted attention.

The floating edible aquatic plant purification system or (FEAPPS) is an ecological engineering system, where aquatic plants float on the water while their roots are suspended beneath the water surface. The fabricated wetland system, being a purification structure for aquatic plants is also well-known. But compared to this technique, plants are more actively managed on FEAPPS and harvested during growth season.

Case study: Characteristics of water purification by Ipomoea aquatica (edible Water Spinach)

The presence of FEAPPS itself inhibits the mixture of sediments since it can diminish waves caused by wind. It can be inferred from the biofilm observed in the roots that the presence of aquatic plants is beneficial to the removal of SS in FEAPPS because of the filtering impact of the roots, or the predation of micro-animals. The average dissolved nitrogen concentrations of FEAPPS plus the control systems can be measured.

Nutrients in the medium-sized FEAPPS were considerably lower compared to those found in the controls system. From spring up to autumn, while the creeping branches of planted morning glory mature actively, the plant takes in nitrogen and phosphorus along the root area while accumulating it in the plant's body. Thus, these results indicate that nutrients dissolved in FEAPPS, like phosphorus and nitrogen, are reduced. Both the phosphorus and nitrogen of installed plant harvested from FEAPPS measured at 4.8% and 0.95%, correspondingly (Department of the Army, 2009).

The average wet weight of water spinach is 16 kg, while the water content is roughly 81%. Thus, the amount of inorganic nutrients removed from harvesting plants is equal to 150 grams of nitrogen and 29 grams of phosphorus. Research shows the turbidity changes of general FEAPPS as well as control systems within different eutrophic aquatic conditions. From research, it can be seen that under all conditions, the turbidity of FEAPPS remains lower compared to that of control systems.

However, the dissolved nitrogen and phosphorus concentrations of the two systems were not significantly different. It's believed that dissolved nutrients are easily mixed with the surrounding water. These factors indicate that the turbidity is composed of microalgae, etc., and is removed by feeding some crustaceans as a filter in the root area. The area is a certain structure, the underwater light intensity is reduced, and plankton growth is controlled by the light control.

Research shows that at each eutrophic level, the concentration of SS, as well as turbidity in FEAPPS, is lesser than the control system since the presence of aquatic plants is beneficial to the removal of SS and turbidity in FEAPPS because of the filtering impact of its roots. In addition, the reason for the extraction of nitrogen or phosphorus by marine plants is for absorbing nitrogen and phosphorus directly through the root area, as well as harvesting plants with nitrogen and phosphorus, plus to exclude the established experimental zone.

The average value of organism biomass and transparency of FEAPPS is much higher than the control system, and the correlation between these materials is considerably high. The root area of aquatic plants has the role of reducing the pressure of predation of plankton since it is hard for most fish to consume plankton. These plankton form an essential part of the water food chain. FEAPPS attracts a number of aquatic animals, like crustaceans and insect larvae (Department of the Army, 2009).

### 4.4.1 Use of Floating Water Plants in Phytoremediation of Heavy Metal Polluted Water

*Source of pollution*: Heavy metal (HM) pollution is regarded as the main source of environmental pollution. Heavy metal contamination in groundwater seriously threatens human health and marine ecosystems. Conventional treatment techniques used to remove contaminants from wastewater are often expensive, time-consuming, harmful to the environment, and most wasteful. Phytoremediation is an affordable green emerging technology that has long-term applicability. The choice of plant species makes the most important feature of successful phytoremediation.

Aquatic plants are highly efficient at removing organic and inorganic contaminants. Hyacinths (*Eichhornia crassipes*), duckweeds (*Lemna minor*) and water lettuce (*Pistia stratiotes*) among other water plants are important metal accumulation plants for remediation of heavy metal contaminated water. By applying innovative methods in phytoremediation, the phytoremediation application of marine plants can be greatly improved. A review of the use of water plants in phytoremediation was collected to introduce the broad application of phytoremediation.

In the past few years, industrialization, agricultural activities, rapid urbanization, and the release of geothermal water as well as olive wastewater, particularly in olive growing areas, have increased polluted wastewater discharged to the environment. Wastewater with soaring pollutant concentrations is extremely harmful to aquatic ecosystems and human health. Wastewater recycling is the only option available to meet the growing needs for water in the emergent industrial and agricultural fields. Industrial and household untreated wastewater have pesticides, oils, cyanide, toxic organics, dyes, phenol, phosphorus, heavy metals and suspended solids.

Heavy metals in these toxic substances can easily accumulate in the nearby environment. Commercial activities like metal processing, automobiles, paper making, mining, geothermal power generation, pesticide manufacturing, dyeing, tanning and electroplating are among the top causes of global heavy metal pollution. It is difficult to remove heavy metals from wastewater since they exist in various chemical forms. Many metals aren't biodegradable; they can effortlessly pass through various nutrient levels to gather permanently within the biota.

Removal of toxic contaminants is extremely important for reducing threats to both human health and the surrounding environment. Heavy metal removal through various technologies (such as ion exchange, reverse osmosis, chemical precipitation, adsorption, or solvent extraction) includes huge operating and maintenance costs plus is generally not environmentally friendly, for these reasons the use of wild remediation techniques like edible water plants is recommended.

These conventional methodologies for heavy metal repair are usually expensive and time-consuming. These treatment technologies need significant amounts of capital investment, which ultimately leads to the issue of sludge disposal. In order to salvage wastewater contaminated with heavy metal pollutants, then an environmentally sustainable and economical treatment technique is required. Current research shows that environmentally friendly phytoremediation expertise can remove pollutants for a long time.

The anthropogenic and geological activities of heavy metals available in the ecosystem are the main sources of heavy metal pollution. Some activities like mining, municipal waste, military activities, urban sewage discharge, fertilizer application, vehicle exhaust, waste incineration, wastewater, fuel production or smelting all produce metal contaminants. A few wild sources for heavy metal pollution are erosion, rock weathering, and volcanic eruptions. The parent material in the weathering process is the main and initial wild source of heavy metals. The use of agricultural pesticides and agricultural soil fertilizers further increases the concentration of Cu, Cd, As and Zn metals in the soil.

The growing demand for agricultural products has increased the use of pesticides, herbicides and fertilizers. Excessive usage of these pesticides may lead to the accumulation of pollutants in vegetation and soil. Sometimes phosphorus and inorganic fertilizers are used to control crop diseases but have the contents of nickel, lead, zinc, cadmium, and chromium at uneven levels.

*The Phytoremediation Strategy:* Phytoremediation is regarded to be an effective, beautiful, cost-effective, and environmentally friendly technique for repairing potentially toxic compounds from the environment. Crops undergoing phytoremediation accumulate pollutants through their roots and then transfer these pollutants to the aerial parts of the human body. The concept of using metal accumulation plants to remove heavy metals and various other pollutants during the phytoremediation process was first proposed in 1983, however, this idea has been implanted for nearly 300 years.

Phytoremediation is known under different names, like agricultural restoration, vegetative restoration, plant restoration, botanical restoration and green technology. The use of vegetation, microbiome and soil, including other agrochemical methods, make plant restoration an attractive green technology that can be used to accumulate different types of metals. The submission of *in-situ* or *ex- situ* remediation is suitable for phytoremediation procedure. *In-situ* application is more common since it reduces the proliferation of pollutants in wastes in water and air, thereby ultimately minimizing the risk to the surrounding environment.

Various contaminants may be treated on-site through phytoremediation, without requiring any disposal sites. By preventing soil erosion and leaching, it also lessens the spread of pollutants. The cost of phytoremediation is much lower compared to traditional remediation techniques, which is the biggest advantage of this methodology. Phytoremediation is a fairly simple technique because it doesn't need any highly specialized staff or special equipment. This applies to the salvaging of large areas, whereby other traditional techniques have proven to be extremely incompetent and expensive.

A large number of pollutants such as pesticides, polycyclic aromatic hydrocarbons or (PAH), chlorinated solvents, petroleum hydrocarbons, polychlorinated biphenyls (PCB), surfactants, explosive elements, radioactive nucleosides, and heavy metals may easily be repaired through the phytoremediation technique. There are various plant species that can accumulate considerably higher amounts of heavy metals in various aspects of the human body (such as stems, leaves, and roots) without revealing any signs of toxicity.

*Features of Phytoremediation Edible Plants:* To make phytoremediation an ecologically sustainable technology, plants must have the following features: wild and fast -growing, high biomass, ingestion of large quantities of heavy metals, then the capacity to transport metals on aboveground plant parts, and mechanisms of tolerance to metal toxicity. Some other factors such as pH, nutrient utilization, solar radiation, and salinity will also greatly affect the plant's phytoremediation potential and growth.

**Figure 4.3:** Some edible aquatic plants like wild rice can be used to decontaminate polluted water.

*Source:* https://midwestfloatingisland.com/hydroponic-phytoremediation/

Phytoremediation follows different mechanisms during plant absorption or accumulation of heavy metals, such as plant extraction, plant stabilization, plant volatilization, and rhizofiltration. Plant extraction, also known as plant accumulation, involves the absorption of heavy metals by the roots of plants and then transferring them to the above-ground parts of the plant where they can't contaminate the water. Once the plant extraction is complete, the aquatic plants can be harvested then burned to obtain energy and metal can be recovered/recycled from the ash as needed.

Sometimes plant pollutant remediation and extraction are applied synonymously, however, this is a misunderstanding. Phytoextraction is a cleaning process, while phytoremediation basically is the concept's name. Plant extraction is a convenient phytoremediation technology for remediation of wastewater, sediments and heavy metals in soil. Plant stabilization meanwhile encompasses the use of aquatic plants to restrict the transfer of pollutants in soil.

The phrase phytostabilization is also called in situ inactivation. Using this technique can effectively repair soil, silt and sediment around water channels. It doesn't however impede the wild environment, plus is a safer alternative. As for phytostabilization, plants inhibit or function as a blockade for water penetration into the soil. When preserving the quality of surface water, groundwater and soil, this technique is ideal for the purpose since it can shorten the movement of pollutants.

Plant stabilization is very effective at treating large sites that are heavily affected by pollutants. It's however just a management method used to inactivate/fix potential harmful pollutants. This is not a fixed solution since only the metal movement is restricted, however, the metal still remains within the soil. On the other hand, rhizofiltration comprises using plants to absorb pollutants, thereby resulting in limited movement of these pollutants in groundwater. Roots also play a very important role in rhizofiltration. Matters such as pH changes in the rhizosphere as well as root exudates contribute to heavy metal precipitation on the root surface.

Once the plant has absorbed all the pollutants, it can be easily collected and disposed of. Root filtering plants often have this ability, to produce a vast root system, gather high levels of heavy metals, simple to handle, and low maintenance costs. Both marine and terrestrial crops with long fiber root systems may be applied for rhizofiltration. The process is used productively to treat agricultural runoff, radioactive pollutants, industrial emissions, and metals. Most of the heavy metals remaining in the soil, such as chromium, nickel, cadmium, lead, copper and zinc, can be properly remediated by rhizofiltration methods.

*Phytovolatilization:* Plant volatilization is a process in which plants convert pollutants into various volatile substances, and then continuously release them into the neighboring environment by means of plant stomata. Some plant species such as canola oil and Indian mustard can be used for plant volatilization of selenium. Meanwhile, mercury and selenium make up the most convenient pollutants which can be remedied from phytovolatilization. Among the top benefits of phytovolatilization is their non-requirement of any extra management when the plantation is finally done.

Other benefits include reducing soil erosion, no interference with the soil, unrequited harvesting plus disposal of crop biomass. The bacteria found in the rhizosphere also contribute to the bioconversion of pollutants, which ultimately increases the rate of plant volatilization. The phytoremediation capacity depends on the plant's uptake ability of various heavy metals found in the soil.

The application of particular chemicals has proven to be an effective technique for increasing the bioaccessibility of heavy metals to crops. Organic fertilizers as well as chelating agents are often used to lower the pH of water or its surrounding soil, ultimately improving the bioavailability and sustainability of bio-accumulation for certain plants. For tobacco, lowering the pH by using chelating agents indicates an increase in Cd accumulation. In various studies, the appliance of ethylenediaminetetraacetic acid (EDTA) supported plant extraction as well as bioaccumulation of Zn, Cd, and Pb.

*Phytoremediation mechanism in aquatic plants:* The aquatic environment is an economical and resource-rich purification system that can be used for phytoremediation in large polluted areas. Aquatic plants are wild absorbers of pollutants and heavy metals. The application of water plants to remove different heavy metals and other pollutants is the most effective and advantageous method. Artificial wetlands and aquatic plants are widely used in wastewater treatment worldwide. The choice of water plant species for heavy metal accumulation is a very significant issue for the enhancement of phytoremediation.

Throughout the years, water plants have gained an increasing reputation for their ability to clean contaminated sites around the world. Water plants always grow a wide range of root systems, which support and makes them ideal for the accumulation of pollutants in roots and branches. The cultivation of water plants is however time-consuming, something that may limit the growing demands for phytoremediation. However, this disadvantage is replaced by many other benefits of this technology in wastewater treatment.

*Importance of Water Plants for Wastewater Phytoremediation:* The phytoremediation of toxic metals with aquatic plants has been widely considered due to its practicality and affordability. Early workers have demonstrated that water plants have the capacity to remove harmful compounds from various wastewaters. Marine plants eliminate heavy metals by absorption or surface adsorption, assimilate them into their systems, before accumulating them in a bounded form. The aquatic plants reduce toxicity in wastewater, making it is less harmful to the immediate environment.

A variety of aquatic plants, such as water hyacinth, lettuce, giant duckweed, *Salvinia* sp., and Manjianghong (*Azolla* sp) are suitable for phytoremediation. They show great ability in various wastewaters.

Phytoremediation of urban wastewater poses a major benefit to the municipal aquatic environment, which is a major cause of heavy metal pollution. Metals such as zinc, copper, nickel, lead and mercury are commonly found in these waters and may cause terminal or acute health problems if left untreated, due to bioaccumulation and phytotoxicity.

The alternative removal methods of arsenic, uranium and boron accumulation were studied for the aquatic plant *L. gibba*. Results indicated that during the initial 2 days of the 1- week experimental research, the plants rapidly absorbed toxic chemicals. In hydroponic research, two massive rooted plants, southern reed and cattail, removed approximately 14 to 85 percent of heavy metals like nickel, iron, copper, zinc, lead, arsenic, magnesium and aluminum from city wastewater. Likewise, marine reeds and cattails have a significant removal rate for heavy metals present in urban sewerage systems. Both these water plants showed higher rates of removal for copper (91%), lead (88%), aluminum (96%) and zinc (85%), and somewhat less removal amount for boron (40%), cobalt (31%) and iron (44%).

The phytoremediation of wastewater and the emission of industrial wastewater into the ground and water means that the threat to people's health, organisms as well as other resources are more serious. Phytoremediation and newly established engineering and biological approaches promote the successful extraction of heavy metals from industrial wastewater using plant stabilization and plant extraction. The phytoremediation ability of 12 kinds of aquatic plants to different heavy metals originating from industrial wastewater sources in the Swabi district of Pakistan was tested. The results show that these marine plants can remove heavy metals from municipal wastewater with exceptional removal efficiency: cadmium (90%), chromium (89%), iron (74.1%), lead (50%), copper (48.3%) and nickel (40.9%).

Southern cattail plants (*Typha domingensi*) have the largest accumulation of iron, zinc, aluminum, and lead, particularly in the roots instead of the leaves of industrial wastewater ponds. It was found that rhizosphere penetration is the main mechanism, which explains the phytoremediation possibility of cattails. Some promising marine plants have better phytoremediation of industrial wastewater than other plants. Compared with the terrestrial plant *Sesbania cannabina*, the aquatic plants *Hydrilla verticillata*, *Marsile aquadrifolia* and *Ipomea aquati* in industrial wastewaters have better accumulation capacity, for the heavy metals or HMs (Fe, As, Hg, Cd Cr, Zn, Pb, and Cu HMs) in industrial wastewater.

Highly diverse aquatic plants have the following restoration advantages: high removal efficiency, improved habitat and distribution adaptability. *T. domingensis* is a highly dominant water plant that's highly tolerant to toxic environments and is good at accumulating HMs. Researchers also report that compared with other highly diversified aquatic plants, the iron, nickel, zinc and chromium released from the wastewater of the metallurgical plant industry have better survival and removal efficiency. This kind of water plant can be applied on a wide scale to study lasting removal performance.

The phytoremediation of textile wastewater is furthermore increasingly popular. Wastewater from the textile industry is regarded to be among the most contaminated wastewater in the industrial sector. Both the dyeing and printing process of textile sector wastewater shall produce both organic and inorganic pollutants. Heavy metals in textile wastewater are more harmful compared to public health systems, and thus more toxic. Researchers studied the feasibility of eradicating bacteria from five different textile businesses in the Lahore region of Pakistan to eliminate copper, chromium and zinc. During the survey, the plant *E. crassipes* effectively removed 94.44% Cu, 96.88% Zn, and 94.78% Cr in industrial wastewater samples.

### 4.4.2 Other Benefits of Aquatic Plants

Research on phytoremediation shows that aquatic plants have certain benefits over other plants with regards to heavy metal restoration. The wide availability, rapid growth capacity, high biomass, cost-effectiveness and toxic pollutant tolerance to make it a suitable accessible phytoremediation plant. The purification systems that use these water plants have attracted increased attention globally because of their ability to accumulate and eliminate persistent organic contaminants from water sources.

In order to absorb and remove heavy metals and harmful compounds in water bodies, the intervention of proper plant remediation methods requires a steady gathering of plant biomass. The adaptation of biomass to exclusive materials is an important factor that promotes the technology's treatment of pollutants. According to various research reports, the aquatic plant biomass can later be applied as animal feed, plus can be applied to generate biogas or compost. Inspection of bio-absorption and bio-accumulation of Lemna's small biomass indicated that it might be used as animal feed as well.

Aquatic plants have sugars in the form of cellulose, starch and hemicellulose. The carbohydrate hydrolysis of fermentable sugar leads to the generation of ethanol, lactic acid, and other essential products. Thus, the sugar found in water plants is a novel and promising feature that supports its role in an ecologically sustainable environment. According to reports, aquatic plants, namely *Eichhornia crassipes* and *Pistia stratiotes*, generate sugar during the enzymatic hydrolysis procedure. Free-floating water plants such as (Spirodela species, Azolla species, Wolffia species and duckweed) may be applied as food sources for waterfowl.

They also offer shelter for small mollusks and insect larvae. Fishes furthermore use these plant mats as cover, plus use their shades to reproduce. Water plants can be effectively used to enhance aquaculture in fish ponds. Eliminating nitrogenous wastes in aquatic plants is another benefit of using them in aquaculture, for example, *Typha angustifolia*, *Echinodorus cordifolius*, *Canna generalis* L., and *Cyperus involucratus* effectively remove ammonia, nitrite and nitrate.

For complete remediation, heavy metals in the environment must be absolutely eliminated as persistent pollutants. The use of phytoremediation is a less damaging, economical and ecofriendly clean technology. Choosing the right plant is the most important aspect of phytoremediation. As mentioned, aquatic plants play a very active role in the removal of heavy metals from polluted sites. Water plants can be effectively applied for bioaccumulation (using living plant biomass) or biosorption (through dead plant biomass) for the eradication of heavy metals.  Interaction, transportation, and chelating agent activity regulate the storage as well as a buildup of heavy metals by aquatic plants.

Genetic engineering improves the buildup and tolerance levels of plants, which indicates that it has a special application in the improvement of phytoremediation effects. For plants, at the subatomic level, different comprehensive steps in favor of transgenic methods have been evaluated to counteract the metal- filled parts of plants. Genetically modified plants have increased tolerance as well as metal absorption ability. Thus, gene manipulation has largely been studied in aquatic plants, however, the genetic modification of aquatic plants for improving their absorption rate of heavy metals HMs is still at the preliminary stage.

Plant biomass disposal, later on, maybe applied for biogas production and may further be applied as general animal feed. The use of marine plants for phytoremediation, the same as otter typical physical or chemical techniques don't need any post-filtration, plus can be efficiently used for treating large volumes of contaminated water or soil. Based on scientific review, the advantages of using water plants for treatment are vast, since this technique doesn't just treat the contaminants, but is also economical and visually appealing, apart from being beneficial for sustainability of entire ecosystems.

## 4.5 SUITABLE CONDITIONS FOR AQUATIC PLANTS TO GROW

**Figure 4.4:** Edible aquatic plants is an aquarium can provide food and nourishment for your fish.

*Source:* https://www.gardeningknowhow.com/ornamental/water-plants/wgen/types-of-aquarium-plants.htm

Like plants growing on land, aquatic plants must have certain basic conditions for survival and to thrive. Same as all living organisms, edible aquatic plants require certain nutrients and environments to flourish. If you want to cultivate a successful planting tank, you need to understand what these things are and try to implement them within your own tank. These live plants can foster wild beauty in aquatic ecosystems, apart from promoting a balanced biome and providing various benefits for your fish, such as producing oxygen and taking in carbon dioxide at daytime, which is good for fish, helps to filter and stabilize pH, prevents algae growth through removal of nitrate and phosphate available in water, and providing a food resource for fish (Tull, 2013).

Whether you just need to add some plants to enhance the ambiance, or build a dedicated outdoor botanical garden, understanding the fundamental needs of water plants can help you capitalize on the success rate and enjoyment of the aquarium. For aquarium plants to flourish, the following conditions must be met: a) clean, fairly soft water b) full-spectrum light c) nutrients and ideal substrates (rooted plants). Without one of the above four factors, all your water plants will suffer. Meanwhile, some of these aspects may also have some negative impact on your plants. To guarantee that your plants have enough light, use full-spectrum lighting of 3 to 5-watts a gallon.

You may further want to apply an automated timer to guarantee that your lights are on for around 12 hours a day. To ensure the plants get the correct nutrition, ensure you use the appropriate substrate in the planting tank and apply fertilizer as needed. Nothing is more striking than a densely planted freshwater reservoir. The key to ensuring the growth and prosperity of water plants is to understand their needs and provide these things. Keep in mind that the needs of different types of plants are slightly different, so make sure to research before storing the tank to ensure that you cultivate the appropriate conditions.

*Water Needs for Aquatic Plants:* The optimal pH value for most aquariums is between 6.5 and 7.8, the hardness is generally between 50 ppm and 100 ppm, and the alkalinity is between 3° to 8° dKH (54 ppm – 140 ppm). Nitrate must be below 10 ppm while phosphate must be under 0.5 ppm to avoid harmful algae from developing on the leaves. The temperature must be between 74 to 80 degrees F. Change 10 percent of the water every week or 25 percent of the water every two weeks and apply organic adsorption resin or reef carbon in the filter to extract organic pollutants which can purify water and lessen light transmission (Tull, 2013).

If your tap water isn't suitable for application with aquatic plants, consider using reverse osmosis and deionized water. Adequate circulation is also essential for plants. It ensures a stable supply of nutrients, impedes algae growth, and prevents organic debris from accumulating on the leaves.

*Light Requirements for Water Plants:* Choosing the appropriate light source for planted aquariums depends on the species to be grown and the height of the aquarium. Some plant species require stronger light to grow, and since light doesn't infiltrate water well, larger aquariums require stronger light sources. Aquatic plants work best in full-spectrum light through a Kelvin score or "color temperature" of 6,500K to 8,000K. Always choose light sources uniquely designed for aquarium plants; lights designed for interior plants don't really have the appropriate Kelvin rating. High-output T5 fluorescent lamps and LEDs provide optimal lighting for aquarium flora.

Many of the popular aquarium models released these days have been designed with more emphasis on interior design, than on the requirements of water plant species throughout the world. People typically think that no matter what plants are in the tanks, the plants will be well. Although sometimes correct, this is not permanently the case. Certain tanks simply do not have the more challenging species. Pairing the right aquatic species to the lighting of your tank is a huge step in the correct direction.

*Fluorescent lighting (T5 and T8):* Fluorescent lamps are readily available, are inexpensive and hardly generate heat, and are very popular. Its main drawback is that it should be replaced around once a year to be effective, and perhaps even every 9 months or thereabout for plants to flourish. In case your fluorescent-tube has been used for at least three years and your plant does not look good, consider replacing it. Basically, there are two key types of fluorescent lights applied in the plant aquarium, which are the T-5 and T-8. Where possible, consider choosing the more advanced T-5 version. In comparison to the older T-8 lamps, they are much better options with numerous benefits.

The diameter of the light bulb is narrower, while the size is only 5/8 inches, not 1 inch of T-8. If you want a large number of planted water tanks, this permits you to install more bulbs in the hood. It is advisable to use a complete spectrum (5000 to 7000K) bulbs. Nevertheless, because of the diverse diameters, T-5 or T-8 tubes aren't interchangeable unless an adapter is used.

*LED aquarium lighting:* LED (Light Emission Diode) aquarium lighting has recently become very popular, and for good reason. They can last for more than 50,000 hours without generating heat, therefore they may be placed rather close to the surface of the water such that your plants can benefit more from the light which they emit. Their tiny and flexible size provides flexibility when placing them. You may also purchase LED lighting in the form of fluorescent tubes instead of fluorescent tubes.

### 4.5.1 Lighting to Avoid in the Plant Aquarium

Always avoid using incandescent bulbs. Unfortunately, some cheaper brand aquariums come standard with these aquariums to reduce costs. They are almost useless for many plants because they simply cannot emit the appropriate type of light which plants need. Where possible, consider replacing them with fluorescent aquarium bulbs. Since fluorescent lamps are cheaper than incandescent lamps to operate, you will save cash in the long term.

No matter which form of lighting you select, your actual purpose must be to illuminate the aquarium for about 8 to 10 hours a day. Applying a timer with a lighting device will ensure that you can achieve every day. If you begin to experience algae growth, consider reducing the time to turn on the light (photoperiod) and observe the situation closely, and adjustments where needed. From personal experience, light intensity offers a better impact on plant growth as compared to photoperiod. Where possible, seek for 4 or up to 5 watts per gallon since this shall ensure improved growth.

Be very careful when applying "watts per gallon" to determine the amount of light a plant requires. The word "watts per gallon" for years has been applied to help the plant aquarium owner to determine the appropriate quantity of light for water plants. Until recently, the formula worked well because standard fluorescent lights were the most typical form of lighting available in the aquarium business. The introduction of high-output T5 or LED lighting has made things a bit more complicated, therefore making "watts per gallon" ineffective.

This is why, bulb power is a measure of the amount of electricity light consumes, rather than the amount of light energy it generates. Light intensity often is measured in terms of lumens rather than watts, and the amount of lumens consumed per watt produced by different light sources is also different. For instance, a 60-watt incandescent bulb generates lesser lumens compared to a 54-watt high-output T5 bulb, plus you don't want to directly watch a 60-watt LED lamp. Lumens, PAR or (photosynthetically active radiation) and PUR (photosynthetic useable radiation) are considerably meaningful measurements.

*Substrate for Water Plants:* First, choose a substrate that promotes root formation in order to ensure that the plant is firmly attached. Nevertheless, when planting an aquarium for the first time, it is best to use sterilized gravel to prevent dangerous bacteria from building up or contamination. However, the sterile mixture also lacks appropriate minerals and trace components, which is very important to ensure that plants grow more appropriately and are adequately supplied for a long time. The use of ready-made mixture concentrates containing minerals and trace components will ensure the proper growth of plants, thereby laying the foundation for healthy and lush plants. It's also recommended that you fertilize the aquatic vegetation periodically.

Choosing the appropriate substrate can ensure that the root system develops normally and fixation of rooting plants occurs. Fine to moderate grade gravel or rough sand is the ideal choice, you can further mix different levels of texture and aesthetic materials. Avoid using ultra-fine sand or coarse gravel since fine sand shall compact and not adequately "breathe", whereas coarse sand prevents proper root anchoring and may gather excessive organic debris. Put up a 2-inch to 3-inch base and tilt it toward the rear of the aquarium. Terraces, hills and valleys may also help form depth and dimension. Try to avoid dolomite, crushed oyster shells, crushed coral, and other calcium carbonate-derived substrates, since they may cause an unwelcome jump in pH levels and alkalinity. Repeatedly rinse the substrate before putting it into your aquarium.

*Aquatic phytonutrients:* Water plants use phosphorus and nitrogen as well as magnesium, manganese, potassium, iron, and other compounds to grow. The majority of the phosphorus and nitrogen come from fish nourishment and waste, however, minerals must also be applied to the aquarium regularly. Based on the plant species, nutrients would be absorbed through roots, leaves or both.

For plants that absorb nutrients through their roots, some water plant enthusiasts blend red earth (an iron-rich clay) into the solution when establishing the aquarium. Others use unique aquatic plant substrates containing embedded nutrients. And yet another technique is to fix fertilizer tablets within the matrix around the plant roots. Don't use plant labels or fertilizer sticks meant for indoor plants, as they don't provide the correct nutritional balance for water plants.

Surprisingly, many people believe that waste produced by fish is sufficient to maintain their water plants, and don't consider adding plant fertilizers. This is a huge mistake, possibly the largest plant-related error that can be made by the aquarist. By failing to add fertilizers, you are fundamentally starving the plants of nutrients they require to survive. Fish excrement, uneaten food, or even tap water can certainly provide some necessary nutrients, but it does not really provide everything you need.

You are likely to encounter issues like leaf holes, discoloration, stunting, and thin leaves. Then eventually they may die. Whichever the case, your water tank won't be like the display water tank you have been seeing online or in regional stores. Same as terrestrial plants, they need macro and micronutrients including carbon dioxide and light. The grower can use various different kinds of fertilizers:

- Substrate fertilizers

In essence, they are almost like soil or made of clay-based compounds, as the name suggests, these fertilizers form part of the aquarium substrate. In fact, you can add these contents only if you set up the aquarium from start. If you plan to create a highly planting tank, you must really consider applying substrate fertilizers. Regardless of their high prices, they can last for many years and provide sufficient nutrition for various strong and healthy water plants. Nearly all the impressive planting tanks you see on the internet will be fitted with bottom fertilizer.

- Tablet fertilizers

Different brands of tablet fertilizers are easily available for use. Nutrient-rich "tablets" may be added instantly to your substrate to provide the nutrients needed for plant species like Cryptocoryne, Vallisneria and Echinodorus. One of their advantages is that they can be fed into the water exclusively when using the plants. Some aquarium owners also add tablet fertilizers when putting new plants into their aquariums.

- Liquid fertilizers

They are the most common form of aquarium fertilizer in today's homes. There are various different products found on the market, plus you must be cautious about finding high-quality products. Generally, liquid fertilizers must be used every week, although check the instructions of each manufacturer, as the instructions for use may vary.

Although they are very suitable for plant species that absorb nutrients directly through their leaves (eg. Egeria densa and Java moss), they cannot provide adequate nutrition for species like the Amazon Sword. Aquarium breeders may have to use a mixture of tablets and liquid fertilizer to feed their respective plants. If you have fish in your aquarium, then consider using liquid fertilizers that do not have nitrogen or phosphorus. These two nutrients shall be present in the fish manure but will be considerably high in the fish tank, where they can cause poor growth of algae.

*Carbon Dioxide (CO2):* Carbon is necessary for the healthy growth of plants. Water plants consume carbon dioxide and produce oxygen in the daytime, and the opposite at night. Many plant lovers add supplemental carbon dioxide at daytime to increase plant size, coloration and growth.

Aquarists at the beginning often ask whether additional carbon dioxide is required for successful water plant growth. The brief answer is no. Many aquatic plants can grow without additional CO2, however their size, color and vitality are often better when added.

Liquid supplements as well as tablets and DIY yeast generators can be used to provide carbon. All of these techniques are effective. However, the most dependable and convenient technique of supplying carbon dioxide is the pressurized injection method harmonized with the water lamp. Since the use of carbon dioxide can considerably increase plant growth, it might be necessary to improve the amount of liquid nutrients added to maintain capacity.

There is no short mention of carbon dioxide or its addition to aquatic plants, or those in existence are incomplete. Plants require carbon just like all other living things and absorb carbon through carbon-dioxide. Some individuals think that in order to make plants grow well, it's necessary to include carbon dioxide in the aquarium. For an aquarium with large amounts of aquarium plants, it may be appropriate due to their higher competition for CO2, or the high ratio of light found in the aquarium, which results in plants that can't meet the photosynthesis needs because of the light amount. However, this isn't often the situation in plants with moderately populated water tanks.

In other words, it is undeniable that the addition of carbon dioxide will promote the growth of water plants. However, adding carbon dioxide is a complex topic, and carbon dioxide shouldn't be added without extra research, since excessive addition may affect your fish or shrimp, which are living alongside the water plants.

### 4.5.2 Aquatic Plant Selection

Choosing the appropriate plant for the aquarium depends on the lighting, the height of the aquarium, visual effects to be created and, to some extent, the kind of fish being farmed. Do research or consult your local aquarium specialist for advice on selecting the best water plant for your tank. When arranging the botanical water garden, place tall or fast-growing plants backward, broad-leaved and "striking" plants at the middle, and short plants in front. Most foreground tree species grow laterally, therefore make sure to leave sufficient space for them to span out. Be careful to avoid planting short plants next to high broad-leaved species, which might block sunlight (Jman Redzic, 2006).

Add some plants to emphasize your aquarium landscape or apply them as the central decorative theme. Some growers have gone to great lengths to build a lush aquatic garden, accented by fish! Live plants input beauty and balance to the aquarium and offer a healthier and wild habitat that your fish can flourish in. Wild marine plants thrive in streams, home aquariums, garden ponds and lakes among a variety of other environments.

However, it is wrong to think that these wild plants can be grown in the aquarium without hesitation. Native wild aquatic plants are usually protected species and are generally not suitable for tropical freshwater aquariums. There is also a risk of introducing bacteria, parasites and algae from ponds and streams into your aquarium. Therefore, you should always buy aquatic plants from aquatic stores. Not only do they have the most attractive plant varieties, but they can also provide the best advice on any special care you need. For example, aquatic retailers will be able to tell you which plants match well (Jman Redzic, 2006).

*Final Thoughts:* Why do people keep plants in the aquarium? In addition to the obvious aesthetic benefits, the edible plant aquarium can also improve water quality within the water tank. Not only do they consume carbon dioxide (CO2) or ammonia (NH4) produced by fish, they also utilize the nutrients required for algae growth, such that they can help reduce or eradicate algae growth in your aquarium. Similarly, they provide cover to help ensure the fish is kept free from stress, providing wild boundaries for their territorial species, and cover for the small fry.

In addition, aquatic edible plants vary in terms of size, shape, and color, and have different requirements for light, feed and water parameters. If you add species that are not suitable for your water habitat setting, then you may have to work harder to keep them in good shape right from the first day.

# 5

## *Wild Edible Plants Found In Forests*

### CONTENTS

## 5.1 INTRODUCTION

What comes into the minds of many when we hear of the word food? To most individuals, food is that which has been prepared by fire, that is, cooked meals. To animals, food is that which they can feed on and be assured that they will be full and in perfect conditions such as having the required strength to carry on with their day to day activities. It is high time, we as human beings, think more like animals and value everything we ingest into our body system. Basically, food should be used to refer to anything consumed by a human being and adds some nutritional value to their bodies. According to this definition, water can be termed as food, this is because water does help in cooling the body, avoiding constipation, it contains mineral salts that boost the immunity of a human being, it helps in food digestion among other multiple health benefits. Fruit juice should also be considered to be food if they are made from fruits. Furthermore, a hungry individual can feed on fruits and be rest assured to acquire enough energy to carry on with their day to day activities normally, without having to think of needing a cooked meal (Press, 2013).

However, it is understandable why human beings refer to cooked meals to be food in most cases. The preparation of a cooked meal incorporates different nutritional components, for example; the preparation of a simple meat stew will consist of tomatoes, potatoes, salt, water, onions, and meat among other components. Furthermore, the prepared stew will not be served alone, it will be served alongside a differently cooked meal, maybe rice, and some chapattis among others. This is the main reason why prepared meals are thought of to be food and no other edible components surrounding our environment.

As discussed earlier, it is high time, we as human beings think of food as anything that is ingested into our bodies' system and adds some nutritional value to us. For instance, in times of prolonged periods of drought, availability of crop products such as the likes of tomatoes, potatoes, onions, animals to be slaughtered and provide us with meat, green vegetables, among other meals which are combined together to prepare meals will be scarce. In such periods, most individuals will opt for other sources of food to boost their immune system and provide them with enough energy to carry on with their day to day activities.

The scarcity of crops introduces us to wild edible plants. Let us go back in times, probably in the first evolutionary age. Did the original man have enough intelligence of preparing a meal, or coming up with fire? The answer is quite obvious to all of us, and it is no. How did this first man survive, for, like the rest of us, he needed strength to carry on with their day to day activities? This brings us to the reality of wild edible plants (Press, 2013).

Without wild edible plants, we would not be in existence today. This is because our ancestors managed to go through life and carry on with their activities such as reproduction all because they were being provided with the necessary nutrients by feeding on wild edible plants. The plants we incorporate with many other crops were originally termed as wild plants, but with time, we turned them into domestic plants. Our main difference with our forefathers is that the intelligence we have currently is more advanced than the intelligence they had back then. The human species is known to have evolved greatly over the ages. Continued evolution means that we discovered new things, new foods, new ideas, and new ways of life, which have made us who we are today.

With evolution, we let go of our old ways, the ways of our forefathers. For instance, in terms of foods, we disregarded the raw meals which our forefathers fed on, for we had discovered fire, we had invented ways of using fire to prepare meals such as meat which our ancestors used to feed on in its raw form. With evolution, we have discovered the sweetness of combining different components to come up with one delicious meal. The sweetness and the luxury of our new discoveries, new delicious meals, have led us to let go, and even to some extent, forget the value of our meals.

It is true that, with the combination of different crops, we are able to come up with one sweet delicious meals. However, we are forgetting that, the real nutritional value of crops is slowly degrading. By feeding on our sweet meals, we are not obtaining the real nutritional value from each of the different crops we are all combining together. Loss of real nutritional value has contributed to the existence of new ailments, ailments which to our forefathers would seem to be unreal, for in their periods, such mysteries like cancer, hypertension among other never existed (Press, 2013).

Such a discovery is slowly taking us back to our old ways. Since most of us barely remember the ways of our forefathers, large investments have been set aside to help in research missions, missions which will take us back to the ways of our forefathers and remind us of our true nature.

Basically, wild edible plants are plants found either in our surroundings or in indigenous forests and are known to be fit for consumption either in their raw of or can be a bit cooked and consumed by human beings. Their consumption is said to be of great importance for scientists believe they are of great nutritional benefit to the human body.

## 5.2 IMPORTANCE OF WILD EDIBLE PLANTS FOUND IN FOREST

As discussed earlier, nations are most times faced by different disasters such as floods, drought, and volcanic eruptions among others. When such disasters occur, most farming activities are greatly affected. For instance, floods take place during the onset of rain, and most crops are known to be setting their foundation growth with the presence of rains. However, with the occurrence of floods, crop growth foundation will be destroyed in one way or the other hence crops will remain stagnant. For the case of a prolonged drought period, a time will come when all the stored harvested crop materials will be scarce, and with minimal rains, the chances of production of new crops will be none. Volcanic eruptions are known to destroy the soil fertility of farmlands reducing the number of farmlands on which food crops are grown. Same to earthquakes, they are known to destroy farmlands. With all these disasters, the availability of the food crops we have been used to will be scarce. However, if we turn our attention to wild edible plants, the chances of them being affected by such disasters are quite low. This means that, if we invest enough energy on wild edible plants as we invest on our grown crops, and get used to relying on wild edible plants for food, the chances of having hunger pandemics in the globe will be very minimal (Angier, 2001).

There are communities known to still hold onto the old ways of their ancestors. However, they are quite rare and barely recognizable for they are remote communities. With the continuous spread of the new adventure most nations are getting into, the adventure of studying wild edible plants, such communities are of great help. Most of such communities are being recognized wild world for they are using their knowledge and value of wild edible plants as their cultural identity in their outside world. In summary, wild edible plants are making some remote and quite indigenous communities to be recognized in the outside world and any needs of they have are being met all because of their value for wild edible plants.

Previously, it has been discussed that most of the cooked meals are not that nutritious for they tend to lose their original nutritional value. This does not mean that they are of no health benefit to their consumers, it just that they are not that rich in terms of nutrients. Furthermore, most grown crops are being produced with the help of genetically modified components which boosts their quantity, their sizes but not their quality. Most of the genetically engineered crops are only beneficial in making profits to the farmer but not adding any nutritional value to their consumer. However, when it comes to wild edible plants, most of them are being consumed in their original form. Furthermore, wild edible plants are wildly grown without any human intervention hence they are rich in wild nutrients as compared to food crops grown by human beings. Apart from being beneficial to the health of human beings, remains from the foods obtained from wild edible plants can be used in boosting the health of our own grown crops. For instance, the chaffs from wild edible cereals could be used as manure in our farms hence boosting soil fertility. Furthermore, such chaffs can also be fed to our domestic animals such as the dairy cattle boosting their immunity and also their productivity rates (Angier, 2001).

Since most governments are slowly acknowledging the importance of wild edible plants, advertisements have been aired in most social media platforms encouraging citizens to value wild edible plants and preserve them. Such a step is being considered to be one among the many plans which have been put in place to help in the eradication of poverty. This is because such advertisements are challenging most individuals who have large acres of land covered with forests to preserve them. the preservation of such indigenous forests is a sure bet that wild edible plants will be present in large numbers hence they can use this opportunity to sell produce from such plants to the government and fellow citizens which boosts their financial stability. Furthermore, preservation of indigenous forests will lower the chances of experiencing low amounts of rainfall hence no more drought, which pushes most individuals to sell their properties leaving them penniless and in total poverty. The presence of trees will help in boosting the soil infiltration process hence, reduce the chances of experiencing floods. With low chances of floods, grown crops will manage to set a firm growth foundation which boosts the availability of enough food for consumption.

Wild edible plants seem to be most favored by the two most common theories known to be used in selection of plants for long term food production. One of the theories is optimal foraging theory which is known to many for supporting the idea that, plants that are sure bet to produce food for prolonged period of time should be the most preferred for growth and preservation. This is because, with such plants, the chances of having food scarcity are minimal, for despite their age, it is a sure bet that they will produce normally. When we compare wild edible plants with our grown food plants, wild edible plants are known to have many years of production before they are considered not fit for more production unlike our grown food crops which are made up of plants known to produce only once in their lifetime. Furthermore, wild edible plants assure us of the fact that, as they continue to age, their overall production rates do increase, the quality of their produce gets better as per their aging period among other advantages (Tardío, J., Pardo-de-Santayana, M., & Morales, R., 2006).

The second theory for long term food production is the theory of non-random plants. This theory supports the idea that only the not random edible plants should be preferred to the others since their uniqueness will enable them to defy most of the challenges they face. As for the food crops we grow in most cases, they are known to be quite random. Their chances of overcoming any obstacles they face are quite minimal for challenges such as drought, prolonged periods of floods among others have been known to destroy them. however, when it comes to wild edible plants, they are not that common in our surroundings, and research has it that they have managed to survive and even have qualitative and quantitative produce at all times despite the challenges they are facing.

Apart from the use of wild edible plants for food purposes, most of them are known to contain some medicinal value. For instance, most of them are known to be more effective in healing most ailments more than the manufactured drugs are. This is because they are wild and unadulterated unlike the manufactured drugs which are incorporated with several artificial products that are of no medicinal value (Tardío, J., Pardo-de-Santayana, M., & Morales, R., 2006).

## 5.3 STUDY OF WILD EDIBLE PLANTS FOUND IN FORESTS

As discussed above, wild edible plants are of great importance to the human race more so at such times when we are facing mysterious problems such as the emergence of incurable ailments, scarcity of food due to food insecurity among other challenges. Furthermore, most wild edible plants still remain to be a mystery to almost all us, for us a large percentage of our current population has lost the knowledge from our forefathers on the identity of wild edible plants and their health importance. With the challenges we are currently facing as a globe, the study of wild edible plants is our only way out.

However, most wild edible plants are facing the danger of extinction. Our current generation has neglected mother nature for long periods. Mother nature has turned her back on us and is slowly taking what we really need now. Indigenous forests known to have wild edible plants in large numbers are slowly drying up. Wild edible plants are becoming extinct. In this study of wild edible plants, governments are making it their first priority the spread of the gospel that, forests need to be conserved, dumping of soily materials in rivers needs to stop, we need to prove to our beautiful mother nature that we care, we need to show her that we value her, and help her give up on her quest of vengeance on us, we need to remind her that we are part of her, remind her that we will take upon our responsibility of taking care of our fellow members of the big mother nature family (Pardo-De-Santayana, M., Tardío, J., & Morales, R., 2005).

In the attempt of protecting mother nature, almost all of the indigenous forests known to be in perfect conditions have been secluded and regarded as national reserves. Setting such forests as reserves will ensure that the chances of destroying such lands are quite minimal hence, protection of the most sought after plants, edible wild plants. Apart from the fact that by preserving indigenous forests we would be convincing mother nature to give up on her quest of punishing us, there lacked a perfect location where botanists would venture as they studied wild plants; that is, apart from indigenous forests, the chances of locating other regions with high numbers of wild edible plants are quite minimal.

Communities around the set aside national reserves have been of great importance in the study of wild edible plants. This is because most indigenous forests are known to be located in the most remote areas around the globe since civilization in such areas has manifested itself in slow rates discouraging the clearing of such forests for the purposes of establishing of more civilized activities. This means that such communities still value their traditions. Botanists and other researchers use all the available data collection methods to try and gather all the necessary information on wild edible plants. Questionnaires are administered to help the researchers understand if the members of the community know some of the available wild edible plants in their surroundings. Samples are collected from these areas for further laboratory study by botanists as they to understand the nutritional value of a certain wild edible plant and the perfect conditions of consumption; that is, should they be cooked or should they be consumed in their wild form. Face to face interviews are conducted to help in facial recognition of the wild edible plants with the help of the communities in such remote areas. This is because only individuals from such communities have the ability to highlight that this is the plant, we harvest this plant organ, we consume either in raw form or prepare it in this manner for consumption, we use for this purpose, and it benefits us in this ways among other valuable information. With this information, the conducting of laboratory experiments seems to be quite easy for the botanists do have a clue of which plant is edible, which organs of the plant are frequently used among other information obtained from the questionnaires and the interviews (Pardo-De-Santayana, M., Tardío, J., & Morales, R., 2005).

After thorough research has been conducted, detailed results are presented to the government. Most of the edible wild plants are scientifically named. Their growth requirements are well defined, their nutritional benefits among other results. The presentation of such information has pushed many governments to venture to the set aside of large tracks of land for more production of such wild edible plants. The lands are set in surroundings with perfect climatic conditions for the survival of such valuable wild edible plants. Well educated personnel is employed for the protection of such plants to ensure that they are in perfect conditions. The use of artificial products such as fertilizers among others is completely prohibited. This is because the use of such products will only lower the wild quality of the plant produce yet what we are all after is the pureness of the wild edible plant produce.

The conducted study has shown that there are thousands of wild edible plant species. Just like normal crops, all of these species have been found in different diversities of the environments. The species belong to different life forms of plant species. The forms of wild edible plants range from grass type to forbs, to shrubs, to actual trees, to climbers among other different plant life forms. Of the different plant life forms, not all of their plant organs are considered edible. Just like in maize, we only consider the maize cone to be edible, the same does apply to wild edible plants. Some of the edible plant organs include the fruits, seeds, tubers, vegetables, and gum among other parts all of which have different nutritional values. The season for harvesting such edible organs differs just like as it differs in our normal food crops (Sundriyal, M., & Sundriyal, D. C., 2001).

## 5.4 LIST OF WILD EDIBLE PLANTS IN FORESTS

There are thousands of edible wild plants found in our forests. Think of a situation where you have gone to a forest hike with friends; and then, you notice that the food you had packed will not last you for the planned duration of your hike. In such situations, many will think of going back for more food. But with the knowledge of wild edible plants, the thought of going back for more food or cutting short your hike will not be an option. Most of the wild edible plants can be consumed even in their raw form. There is no need of making a fire to cook them.

*Fireweed:* The fireweed is a forest plant scientifically known as the *Chamerion angustifolium*. It was named so after the scientists who discovered it. It is known to many because of its conspicuous pink flowers which have their male flower part, the stame ns, loosely hanging and exposed by the corolla. This is because the corolla is open and not tightly enclosing the internal parts of a flower. The stems of the fireweed are quite smooth and straight with a shiny reddish color that is visible from far. Their leaves are quite unique in that they have a circular vein patter that has no visible ending from the leaf margin.

**Figure 5.1:** Image of the fireweed plant.

*Source: https://matteroftrust.org/62-edible-wild-plants-that-you-didnt-know-you-can-eat/*

To most people, the firewood is referred to as the rosebay willowherb in Britain. The edible parts of the fireweed plant are the corolla flower parts, the unique leaves which are said to have a sweet juice in them and the smooth coverings of the stem can be pilled off and chewed. Some individuals sip the sweet nectar fluid from the nectaries of the flower (Angier, 2001).

*Dandelion:* The dandelion is a simple forest edible plant. Its entire structure resembles the structure of the sunflower plants more so in the fact that they have yellow flowers whose shape resembles the flowers of the sunflower plant, however, they belong to the same plant family. The dandelion grows in forests experiencing little high temperatures most of the time and moderate rainfall.

To most people, the dandelion does not seem like an edible plant, but it is actually edible. All its plant organs are edible, meaning that the flowers, leaves, roots, stem, the entire plant is safe for consumption. Some individuals may choose to cook the leaves and the roots, but whether cooked or not, they are fit for consumption and do enrich the human body with necessary nutrients.

*Chickweed:* The chickweed plant is scientifically referred to as the *Stellaria media* plant, named after the scientists who discovered it. It is a bushier plant with beautiful white flowers that surround the entire plant structure. It is a more advantageous wild edible plant in that it can be used as both a healing herbal more so for cooling purposes to individuals with fever issues and it can also be used as a nutritious edible plant.

They are easily noticeable to many because of their fine hairs which are located on only one side of their stem, which grow in visible straight lines. The chickweed plant has fine hairs on the sepals; that is the green support organs of their flowers which attach them to the plant branches.

Their leaves and flowers are the most edible plant organs of the chickweed plant. In fact, the leaves can be used as ingredients in the preparation of prime salad (Angier, 2001).

*Curly Dock:* The curly dock is a wild edible plant scientifically known as the *Rumex crispus*. It has tiny light green leaves, which grow close to each other on top of each other alongside the tree branches. It not that much of a healthy edible plant for it does have some effects, though they are not that serious for they do not require any type of medication for them to heal. It is only safe for consumption as a snack, hence it should be consumed in small quantities. The effects stated earlier include urinary tract irritations which last only for a limited duration after their consumption. The leaves of the plant are the ones mostly fed on. They are consumed in their raw form, hence no need for setting up a fireplace to cook them. Apart from being consumed as snacks, the leaves are quite appetizing in that they increase the urge for more food. The leaves are the main source of both vitamin A and vitamin B hence they boost the immunity of the human body.

*Asparagus:* The Asparagus is a wild edible plant known to many people because of their fleshy green spears which can be identified from far. They are quite advantageous in that they can be consumed in either their raw form or be boiled before consumption. They are quite nutritious for they are known to be rich in vitamin C and potassium.

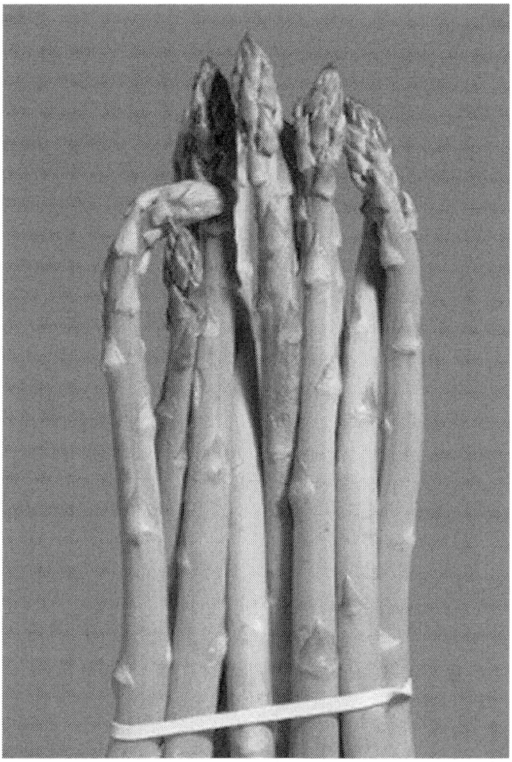

**Figure 5.2:** Image of the Asparagus plant.

*Source; https://matteroftrust.org/62-edible-wild-plants-that-you-didnt-know-you-can-eat/*

Since the Asparagus is a wild plant, it is known to have thinner stalk sticks. Nowadays, the stalks are being harvested and sold in our grocery stores. The stalks are said to be fit for harvest once they are approximately six to eight inches in terms of width and approximately three to six centimeters in terms of their height. During consumption, they should be washed thoroughly and both the top and bottom parts chopped off if they are to be consumed in their raw form; however, if they are to be boiled before consumption, still both the brownish top and bottom parts need to be chopped off and the rest of the stalk diced and boiled in a medium amount of water.

*Chicory:* The chicory plant is a wild edible plant that belongs to the same plant family as the dandelion plant. It is known to many because it is a very beautiful and attractive plant from a far more so because of its conspicuous bright blue flowers. It is more of a busy plant in that it grows as a tiny bush.

All parts of the chicory plants are edible, however, their form of consumption different from one organ to another. For instance, the leaves and flowers can be consumed in their raw form, but the roots need to be boiled before consumption. The roots must be thoroughly cleaned, their outer layer peeled off and then boiled in medium water for a certain duration before consumption.

*Wood Sorrel:* The wood sorrel is a wild edible plant that grows in forests with warm temperatures and high amounts of rainfall. It is a beautiful plant for it has ovate shaped leaves which grow in threes in one leaf branch. The flowers are white in color with both the female and male parts of the flowers; that is the carpel and the stamen respectively, being located deep in the flower. The top parts of the corolla part of the flower are quite loose and hanging on both sides in fives for the flower has five floral parts.

The color of the flowers does change from time to time depending on the maturity level. For instance, young flowers are known to be white in color; medium flowers, flowers which are not yet ready for reproduction are green in color; and the flowers ready for reproduction is bright yellow in color. The yellow color is quite conspicuous for the purpose of attracting insects which will help in pollination. The yellow mature flowers do have sweet nectar fluid so that the insects may feed on it as the anthers get attached to the insect body or are deposited on the stigma of the flower.

**Figure 5.3:** Image of the wood sorrel.

*Source;* https://matteroftrust.org/62-edible-wild-plants-that-you-didnt-know-you-can-eat/

The wood Sorrel is entirely edible in that all organs of the plant are edible. The leaves can be consumed in their raw form and they are said to be a great source of vitamin C. The flowers, which are consumed in their raw form, the roots which can be boiled among other parts are known to have a bit of a sour taste like the lemons. The entire wood sorrel is rich in vitamin C.

*Bull Thistle:* The bull thistle plant, as the name does somehow suggest is a thorny wild edible plant. It is dominant in quite dry forests. Almost the entire plant is thorny. The leaves do have sharp and quite extended thorns. The stem has firm thorns. Only the root is known to lack thorns. The thorns help the plant in protecting itself from its predators. The thorns do also help the plant in survival in dry surroundings. This is because the leaves act more as the leaf stomata hence reducing the amount of water which the plant transpires for they will ensure transpiration takes place in the form of minute droplets and not big droplets.

Due to its unique structure of being covered with thorns almost in all plant organs, most individuals consider this plant to not be fit for consumption. The roots and the leaves of the bull thistle plant are known to be safe for consumption both in raw and boiled form. However, before consumption, the thorns on the leaves must be removed to prevent them from hurting the consumer in any way.

*Alfalfa:* The alfalfa plant is a wild edible plant that grows in forests in in warm climatic conditions. It belongs to the pea plant family. The plant is a flowering plant, which has conspicuous purple- colored flowers. The roots of this plant are known to grow to great depths hence firmly supporting the plant for it grows to greater heights. Apart from the support purposes, the roots have to go deep into the ground to ensure that they absorb an adequate amount of water from the underground water levels so as to support the growth of the plant by supplying it with enough water for its growth activities.

**Figure 5.4:** Image of the alfalfa plant.

*Source; https://matteroftrust.org/62-edible-wild-plants-that-you-didnt-know-you-can-eat/*

Alfalfa is a wild plant which can be used as either a medicinal plant or an edible plant. As a medicinal plant, the roots of the plant are boiled and used in the treatment of alcoholism and reduction of drug dependency more so for drug addicts. As an edible plant, the plant is said to be rich in different vitamins. Not all organs of the alfalfa plant are edible, only the leaves can be consumed in their raw form as the roots must be boiled before consumption.

*Broadleaf Plantain:* The broadleaf plantain is a wild edible plant known to be dominant in forests found in dry environments. As the name suggests, the broadleaf plant belongs to the plant family of the plantains. It has shiny green leaves that are oval- shaped. The stem to the leaves is quite thick and all leaf stems meet at the base of the plant. The plant is not that tall for it is approximated to be at least three to five centimeters long.

The plant has been categorized as a wild edible plant, however, not all organs of the plant are edible. For instance, only the leaves can be considered to be edible in raw form and only in while they are young and tender. Generally, the broadleaf plantain plant is known to be rich in vitamin A, vitamin C and vitamin K.

*Creeping Charlie:* The creeping Charlie was basically named after the scientists who discovered it hence its last name, Charlie. The plant is not that popular for it is quite tiny in size with uniquely shaped bright green leaves.

Almost all organs of this plant are covered with fine hair, including the roots. The roots are quite thick. Since it has been categorized as an edible plant, only the leaves are known to be edible and in their raw form, that is, they do not have to be cooked by boiling or other means. Generally, the creeping Charlie plant is considered as a rich source of vitamin C; for centuries it has been recommended to many individuals suffering from different ailments caused by lack of enough nutrition in their body.

*Forget me not:* The forget me not is a wild edible plant scientifically known as the Myosotis. Most individuals think it was given the name forget me not since it is quite attractive and the chances of ignoring it once you come across it are quite minimal. It has tiny flowers that are tender and have a conspicuous blue color. This plant is more visible during the spring season since by then the temperatures are quite suitable for its survival.

**Figure 5.5:** Image of the forget me not plant.

*Source: https://matteroftrust.org/62-edible-wild-plants-that-you-didnt-know-you-can-eat/*

The flowers are the only known organs of the plant which are safe for consumption. They are consumed in their raw form. Since the flowers are minute, they can only be consumed as snacks and not as proper meals.

*Garlic Mustard:* Garlic mustard was named garlic because it naturally emits a strong odor just like the likes of the garlic onion. The plant produces flowers in the month of June in most regions. The flowers are small in shape with a creamy color which makes them quite attractive from a distance. The leaves of this plant are known to be quite broad, kidney shaped leaves. The plant produces slender seed pods which contain very tiny seeds. Almost all organs of the garlic mustard plant are edible; from the produced seeds which can be consumed in their raw form or be boiled, to the leaves, the minute flowers among others. Generally, the plant is said to be quite nutritious.

*Wild black Cherry:* The wild black cherry is a forest edible plant that is scientifically known as the *Prunus serotina*. It has medium -sized leaves which a light green in color and as the name suggests, produces black colored cherry fruits. Since it is an edible plant, almost all parts of the plant are edible. The fruits have tiny seeds that can be extracted and converted into cyanide becomes the seeds contain a compound that is rich in cyanide. The fruits are also edible; however, it is advisable that they be consumed only in small amounts. The plant is quite nutritious.

*Harebell:* The harebell is a forest edible plant that is quite small in size for its height ranges from around thirty centimeters to sixty centimeters long. It has slender stems known to produce a white sap once cut or the outer layer peeled off. The plant has blue colored flowers which are quite sagging. Only the leaves are known to be fit for consumption. They can be consumed either in their raw form or be boiled before consumption. The leaves are said to be rich in Vitamin C.

*Elderberry:* The elderberry is wild edible plant, which is dominant in forests located in either temperate regions or subtropical regions. In most parts of the world, the elderberry plant is used as a raw material in the manufacture of ornaments. The plant produces fruits that are black in color, the berries. The leaves are long and slender. The leaf branches are seen to be pink in color with the rest of the plant stem being light pink in color.

Apart from the use of the elderberry plant for ornaments, it is also an edible plant. The flowers, fruits and the leaves are all edible more so in their raw form, however, the leaves are mostly boiled before consumption. Raw berry fruit does not have a great taste hence it has to be boiled before consumption or wait until it is ripe for consumption.

*Field Pennycress:* The field pennycress plant is known to have a life span of only one year, in that it only survives annually. It is said to produce round, flat and a bit winged seed pods that have a deep apical notch. The plant has a height of approximately sixty centimeters. The plant has been ruled out to be an edible plant, though research has it that it is not that beneficial. Only the leaves are edible. They can be consumed either in raw form or boiled for consumption. Unfortunately, the leaves are not that tasty and they have conditions for them to be harvested; that is, they should be harvested before the flowering season of this plant. The flowers are creamy- white in color.

*Coneflower:* The coneflower is a wild edible plant that belongs to the same plant family as the sunflower plant. This means that the plant grows in hot climatic conditions for it is able to resist heat and drought. The flowers of this plant resemble the flowers of the sunflower plant only that their color varies depending on their maturity level. The flowers are purple colored while young and once mature they change their color to pink. The flowers display a prickly seed cone at the center just like the sunflower. The leaves are known to have a rough texture due to the high temperatures they have to endure.

Only the leaves and the flowers are edible. Both of them are edible in their raw form. As for the flowers, only the matured pink flowers are safe for consumption.

*Kudzu:* The kudzu plant is a forest plant that is more of a vine plant. The vine of this plant does have a yellow-green color. The leaves are quite large and they do shed annually more so during the summer season in an attempt to reduce the rate of transpiration hence conservation of enough water in the plant for different growth activities more so for survival. The plant is entirely edible in that almost all of its organs are safe for consumption and they have been ruled out to be quite nutritious.

*Meadowsweet:* The meadowsweet plant is a wild edible plant of the "royal class" of plants since it is found in the most beautiful surroundings in forests grown in hot and wet regions. The plant is commonly referred by as the queen of the meadow or the lady of the meadow or the bridewort plant. The Bridewort plant is quite advantageous in that it has two benefits. One of the benefits is that it can be used as herbal medicine to heal different ailments and the second benefit is that it can also be used as an edible plant. However, for the edible purposes, only the leaves of the meadowsweet plant have been ruled out to be safe for consumption, more so in their raw form. The leaves can also be boiled for consumption though their nutritional value will drop due to the effects of fire. The roots are boiled or used in the formation of decoction which can be used to heal different ailments such as the likes of cough, relieve asthmatic effects among others. Meadowsweet plant is quite a large tree that grows in the form of a clump-forming plant.

*Mallow:* Mallow tree is a forest plant scientifically referred to as the Malva. It is said to be one of the earliest scientifically discovered edible plants. Since it one of the earliest discovered forest plants to be edible, it had the privilege of being given an English name, the Mallow name, because it is rare to find a plant that has an English name. It is known to have a quite long and really deep fleshy taproot. The root is that deep to ensure that there is enough supply of water to the upper parts of the plant since the Mallow tree is known to be dominant in areas that experience water shortages. The plant has light purple flowers whose inner floral parts are a bit exposed due to the loosely hanging corolla parts of the flower. The leaves on the other hand are a bit slender, and more of lanceolate in shape with a dark green color.

**Figure 5.6:** Image of the mallow plant.

*Source: https://matteroftrust.org/62-edible-wild-plants-that-you-didnt-know-you-can-eat/*

Since it has been ruled out to an edible plant, almost all organs of this plant; be it the roots, the stem, the leaves or the flowers, are safe for consumption. The leaves can be used in the preparation of lettuce stew among other kinds of stews. Generally, the mallow plant is entirely known to be rich in different types of vitamins and mineral salts such as the likes of vitamin A, vitamin C, calcium and iron, respectively. The vitamins and mineral salts depend on the plant organ which has been consumed, for as stated earlier, the plant is entirely edible.

*Peppergrass:* As the name suggests, the peppergrass is a type of forest grass that grows in cool and wet forests. It is completely a green -colored grass that produces quite beautiful minute white flowers. Peppergrass is known to have a height ranging from fifteen to sixty centimeters depending on the rain intensity of the region it has grown. Despite the fact that it is of the grass type, it has a quite strong stem from which tiny branches originate from. The grass produces seed pods that are used for different purposes. Mostly, the seed pods are used as edibles since they are safe for consumption. The seedpods are used as a substitute for black pepper, meaning they are quite useful in the adding of flavor in our prepared meals.

*Pineapple Weed:* The pineapple weed plant is quite a beautiful wild plant. It has conspicuous flowers whose color is creamy white with an ovate shape. It produces fruits which in the shape of the actual pineapple fruit hence, its name the pineapple weed plant. The plant belongs to the same plant family as the chamomile plant hence their great similarity for in most cases people confuse the pineapple weed plant for the chamomile plant. The only difference between the two plants is said to be in the flowers; that is the flowers of the pineapple weed plant have six floral parts while the flowers of the chamomile plant have three to four floral plants. The plant is known to grow to a maximum height of approximately thirty centimeters tall.

The pineapple weed plant is an edible plant, and as the name suggests, it does have quite delicious edible plant parts. Only the flowers and the leaves are safe for consumption, more so in their raw form, and their taste is said to be really appetizing. However, the flowers could be boiled for consumption though their nutritional value will be a bit low. The plant is mostly fed on by hikers cause it can only be consumed as snacks since both the flowers and the leaves are quite tiny and for one to feed on them and be full, it will take time.

*Pickerelweed:* The pickerelweed plant is a wild plant known to dominate in watery surroundings like the shallow freshwater in lakes, streams, rivers and even swamps. It is quite a long plant with a thin stem. The leaves mostly grow on the top part of the plant. The leaves have a glossy green color and they have a parallel leaf venation. The flowers are violet-blue in color and the plant is known to have its flowering period during the summertime, for the winds are quite high, insects are busy looking for food hence, the agents of pollination will be readily available giving the plant a chance to reproduce. The flowers produce tiny seed pods that contain tiny seeds inside them.

Since it has been ruled out to be an edible plant, the leaves, flowers and the produced seeds have been said to be safe for consumption. The flowers, seeds and leaves can be consumed in their raw form, however, for the seeds, they are bit sweeter once roasted.

*Mullein:* Mullein is known to be an attractive wild plant which has some tender, velvet- like structure. It is quite tall for its height ranged from one and half a meter to two meters. It has a straight stem. Just like the pickerelweed plant, the leaves and the flowers of the mullein plant grow at the top part of the mullein plant. It has light yellow flowers which are quite beautiful from far. The flowers are reddish in color and they are quite tiny. As for the leaves, they are minute in size and of the ovate leaf shape.

**Figure 5.7:** Image of the mullein plant.

*Source: https://matteroftrust.org/62-edible-wild-plants-that-you-didnt-know-you-can-eat/*

The mullein is an edible wild plant hence most of its plant organs are edible. The leaves and the flowers are said to be safe for consumption, however, they can only be fed on as snacks since they are small in size and it might take eternity before one consumes them to satisfaction of their hunger. The leaves and flowers are not that much of a great- tasting snacks, hence it is advisable that they be used in the preparation of tea. Still, their use as beverages does have some nutritional value hence boosting the immunity of the human body.

*Red clover:* The red clover is a wild plant found mostly in hot and wet forests. The plant is not that tall for its height ranges around seventy to one-meter-tall since their stem is not that erect. The leaves of the red clover are of medium size with a lanceolate shape. The flowers are somehow reddish in color hence the name red clover plant. They appear in the form of a cluster of flowers and they are quite conspicuous, a feature that attracts insects that help in flower pollination for plant reproduction purposes. The cluster of flowers is made up of different tubular-shaped flowers. The red clover is a wild edible plant hence most of its plant parts are safe for consumption. The leaves are said to be safe for consumption even in raw form but as for the flowers their edibility still remains to be a mystery for some communities say they are safe for consumption, others say they are poisonous, and as for the scientist, they are yet to clarify this fact. The flowers produce clovers which are said to be the sweetest clovers of all other clover types of all clover kind of plants.

*Partridgeberry:* Partridgeberry is a rare wild plant found in only a number of indigenous forests. It is a short plant of a height approximately ranging from twenty to fifty centimeters high. The stem is not that straight and it is known to have different colors depending on the season. In winter seasons, the stem has a light green color but in summer season the stem has a light brown color. The plant produces quite fuzzy flowers, which later produce red colored berries. Since the partridgeberry is an edible plant, only the leaves and the berries are safe for consumption. However, the sad truth is that both the leaves and the berries, do not have that much of great taste. Most of their consumers consume them due to lack of a different kind of food or because of the nutritional value, they add into the body for they are known to be rich sources of different nutrients.

*Sheep sorrel:* Sheep sorrel is a wild plant that belongs to the same plant family as the blueberry plant. This is because they grow in regions having the same climatic conditions. The plant is dominant in forests which are mostly covered with sandy soil and also covered mostly in the grass than fellow trees. The advantage of this soil to them is that it has great drainage ability, hence in summer seasons, the plant is able to obtain water from the underground level for it had been well preserved around there. It is said to be approximately sixty centimeters high. It has pink flowers which produce berry- like fruits, which are red in color. It has a minute leaves that are slender in shape and a bit of shiny green in color. The flowers tend to grow mostly on the upper part of the plant where they are exposed to free space allowing for pollination through the wind agent of pollination. Some of the sheep sorrel flowers are pollinated with the help of insects. The plant does have a straight stem which is said to have a bright red color such that it is visible from far.

Since the plant is an edible plant, it leaves are safe for consumption either in their raw form a slightly boiled. As for the other plant parts, they are yet to be ruled out as safe for consumption.

*Shepherd's purse:* The shepherd's purse plant is quite a rare and unique plant. First of all, it has a unique structure and a unique name. The shepherd's purse name was derived from its fruit structure. The fruits are known to have a triangular shape and they are flat hence having a purse-like appearance. Secondly, the plant is unique because of its flowering capability. Most plants are known to have either one or two flowering seasons, but the shepherd's purse plant, it always has flowers throughout the year; that is, it does not have specific flowering seasons. Its flowers are creamy -white in color and of medium size for they are somehow visible from far. As for the leaves, they are also medium- sized but are more of a curly shaped and have a very light green color.

The plant has been shown to be in the category of wild edible plants. The leaves are said to be safe for consumption and they are a great source of few nutrients.

*Sunflower:* The sunflower is not that much of a wild plant, but it is sometimes categorized as a wild plant because it does grow wildly in forests without the intervention of human beings. The plant is scientifically known as *Helianthus annuus*. The sunflower plants are said to grow to a height ranging from one and half a meter to around two meters high. The flowers of this plant are bright yellow in color and they are quite conspicuous so as to attract insects that pollinate them. These flowers can only be pollinated with the help of insects for their anthers are quite sticky hence can be transported from one plant to another only with the help of insects as their carriers.

To most individuals, the sunflowers are known for their use as raw material in the production of oils more so the cooking oil. However, since it was originally a wild plant, and to some communities, it does still remain to be a wild plant, it has been discovered to be edible. The flowers to the sunflowers produce tiny seeds that are safe for consumption in their raw form and are quite nutritious. Additionally, there is a bit of a difference between the commercially grown sunflower plant and the wild sunflower plant. The difference is that the wild sunflower plant produces tiny and quite delicious seeds while as for the commercially grown sunflower plant, it is known to produce a bit bigger seeds, which are not that sweet. The difference can be easily explained in that; most of the commercially grown sunflower plants are incorporated with artificial compounds during their growth which boosts the quality of their end products, unlike the wild sunflower plants which have to persevere different challenges for them to strive to the point of production.

*Spring beauty:* Spring beauty plant as the name suggest is a very beautiful wild plant because of its flowers. They are scientifically referred to as the *Claytonia carolinianais*. Its flowers are of medium size and they are very conspicuous from far because they have pink petals that have veins of dark pink in color. The young flowers are white in color with the mature flowers being pink in color.

The plant has some nutritional value in some of its plant organs. The roots, leaves and the stem part of the spring beauty plant have been shown to be safe for consumption.

*Tea plant:* The tea plant can also be categorized as a wild edible plant because of two main reasons. One of the reasons is the fact that this plant was originally a wild plant before some communities chose to plant it as a cash crop. The second is that to most people, the tea plant is known to only have one benefit, and that is its use in the preparation of beverages; however, surprisingly the tea plant does have some parts which are safe for consumption in their raw form. The plant is known to last for approximately four to five decades, still in perfect condition and producing produce of quite good quality. The flowers of the tea plant are known to be white in color and they do have a special feature of producing sweet fragrance more so in the onset of their blossoming season. The flowers are pollinated by insects more so the bees for they do produce sweet nectar fluids that are helpful to the bees in the manufacture of honey.

For the edible purposes, the same tea leaves used in the manufacture of tea are used in preparing different beverages, are safe for consumption in their raw form and quite nutritious. The flowers are also safe for consumption in their raw form and they also do have some nutritional benefit.

*Toothwort:* The toothwort is a wild plant belonging to the same plant family as the garlic mustard plant family. It is known to dominate in forests rich in thick soils such as the likes of clay soil. The plant does have long and slender leaves which are of the lanceolate leaf type. The flowers are shiny white in color having approximately four floral parts. However, the color of the flowers does vary depending on size, for example; while young some flowers are known to be pinking in nature; at the medium growth stage they turn to light purple and at the maturity level they flourish into the known beautiful white flowers.

**Figure 5.8:** Image of the toothwort plant.

*Source; https://matteroftrust.org/62-edible-wild-plants-that-you-didnt-know-you-can-eat/*

Toothwort is an edible plant despite the ugly name which to most people convince them that they are not at all edible. However, not all plant organs of the toothwort are edible. Only the leaves and the roots are known to be safe for consumption in their raw form. The roots may be boiled a bit to enhance the taste though the boiling does lower their nutritional content in them.

*Teasel:* Teasel plant is a quite complicated wild plant. This is because most of its plant parts are covered with sharp hairs. It grows in dry forests hence the presence of the sharp hairs which reduce the rate of transpiration for the plant making it possible for the plant to survive even in very hot periods. The plant reproduces through the flowering process. Its flowers only blossom during the summer period; and for them to come out, they first form a ring -like structure around the tip of the plant branch.

The plant produces thousands of seeds in one reproduction period. The seeds are a source of food to the birds of the air and other wild animals. Sometimes the seeds fall on the ground and they enhance the fertility of the soil upon decay. The teasel is not that much of an edible plant for only its leaves are safe for consumption; and moreover, the leaves can only be consumed at their young stage of growth and once the sharp hairs covering them have been removed.

*Wild grape vine:* A wild grape vine is a wild plant that has some unique structural characteristics in that it lacks a solid upright straight tree stem like most vining plants. The plant grows quite high in its own ways and it has a thick stem. The leaves are lobed in shape with a dark green color. They also have a sharp end and a very smooth texture on both sides of the leaf.

Since the plant has been ruled out to be an edible plant, only its leaves are safe for consumption. However, the leaves can only be consumed as snacks since they are not that big enough to be able to completely satisfy the hunger of anyone.

*Wild bee balm:* The wild bee balm plant is a wild plant known to grow in dry and sunny areas. The plant was given the name wild bee balm because it produces quite beautiful flowers which are dominated by bee insects. The flowers are thick lavender flowers with a dark pink color. They are known to produce sweet nectar juice which is used by bees in the manufacture of honey. Furthermore, the flowers are known to produce a sweet fragrance throughout their blossoming period. The leaves of this plant are medium-sized and they ovate in shape. Only the leaves of the wild bee balm plant are safe for consumption, however, they are consumed as snacks.

*Vervain mallow:* The Vervain mallow plant is wild plant that dominates in forests found in hot temperature regions that have thickets and are more of waste drylands. They are known to have a height that ranges from around fifty centimeters to one meter. They have small -sized leaves that are light green in color. Their flowers have four to five floral parts and their color does change depending on the maturity level of the flower. The young flowers are white in color while the mature flowers are light pink in color.

The plant is a wild edible plant hence its leaves, flowers and seeds are safe for consumption. However, all of them are only recommended as safe for consumption in their raw form.

*Prickly pear cactus:* The prickly pear cactus is a wild plant known to dominate in regions that experience high temperatures. It has white flowers which are minute in size. It does produce fruits that are green in color and either red or purple once ripe. The fruits are of the shape of pear fruits. Almost all parts of this plant are covered by sharp thorns which lower the rate of transpiration since the plant is known to dominate in regions that experience low amounts of rainfall.

**Figure 5.9:** Image of the prickly pear cactus plant.

*Source; https://matteroftrust.org/62-edible-wild-plants-that-you-didnt-know-you-can-eat/*

The prickly pear cactus is quite a beneficial plant for it can be used as both a medicinal plant and an edible plant. For medicinal purposes, its roots are used in the preparation of a decoction known to heal different ailments such as the likes of cough, bronchitis, asthma among others. As for food purposes, the fruits are safe for consumption once ripe, however, the sharp spines have to be removed carefully for they might be harmful to the human hand, before consuming the fruit.

# 6

## *Identifying Common Poisonous Wild Plants*

### CONTENTS

## 6.1 INTRODUCTION

Not all plants have medicinal value as well as nutritional value, there are plants that mainly produce toxins which are specifically meant to prevent any organism from feeding on it. Plants are not like animals who have the ability to escape danger as well as protect themselves, they depend on these toxins to protect them against living organisms. There are those plants that contain parts that help to protect them and those parts can include, thorns, spines as well as prickles but the most common form of defense in all of them is their chemical properties that is the toxins they produce.

Plants have undergone a lot of evolution and they have developed more complex ways of ensuring that they are protected against other organisms that can feed on them. They have been able to choose chemicals that help in the protection process, chemicals that offer the best protection. One of the compounds is known as tannin that is very toxic when consumed by other organisms and it has also proved to be fatal in some cases. Tannins came up way before other compounds were found, they were among the first compounds that helped in the protection of plants (Ballerini, 2012).

The younger plants that are still developing or those that have developed and are small in stature contain a compound known as polyacetylenes and an example of these plants is Asterales. These compounds play a major role in ensuring that the safety of the plants is met at all times, it prevents the plant from being consumed by human beings as well as animals. If any of them tries to eat the plant then it might have some very serious side effects as well as they could die from poisoning.

In as much as these plants have toxins that may prove to be fatal, they can also be used to cure certain illnesses as well as prevent and control some illnesses if and when they are taken with absolute caution as well as under the supervision of a medical practitioner. There are many chemical compounds in plants that protect the plant and a lot of questions in relation to these compounds have come up and they include the specific plants that contain this defense mechanism, the living organisms that the plants are usually protected from, the level of toxicity in these compounds as well as the medicinal benefits of these plants.

These questions have led to a lot of research especially in the field of ethnobotany and as a result, people have learnt to understand plants better as well as know which type of plants are safe to consume and which ones are not. Often there is a lot of misunderstanding on the difference between plants that have psychotropic effects as well as properties and those plants that have toxins that are poisonous. The psychotropic plants when taken in large amounts have proven to very fatal in as much as they are safe for human consumption (Ballerini, 2012).

Plants that are considered to be poisonous actually have toxins and produce toxic compounds which can as well cause, infections caused by bacteria, infections that are as a result of a virus as well as those infections that are caused by parasites that are considered to be fungal. A plant like the peanut plant, it can be poisonous to those people who are allergic to peanuts because of the compounds that it produces.

The following are some of the categories of poisonous plants with their examples.

## 6.2 POISONOUS FOOD PLANTS

Plants can be used as a source of food and it is possible for these plants to contain toxic compounds, these toxic compounds might only be present when the plant is of a certain age and they can be eliminated if the plant undergoes processing for the purpose of removing these toxins. These plants in as much as they all have toxins, some of them are only poisonous to some people like children, the old people as well as those people who have underlying medical conditions. They can also be toxic to some animals and not all, they can be poisonous to either dogs, cats or other forms of livestock. A lot of these plants with nutritional value can be consumed by people who have good immune systems but they need to exercise a lot of caution (Nyerges, 2016).

Some of these food plants include:

*Onions and garlic:* Onions are known as *Allium* spp scientifically and they belong to a family of plants known as Liliaceae. This plant mainly grows in areas that have a lot of bushes as well as thickets but they can as well be found in hills where the ground is bare, as well as areas that have a lot of sand in the low areas of the land. The animals that are mainly affected by these plants include: cattle and cats whereby these are usually affected and also the plant can be lethal to them, horses and dogs, on the other hand, can be affected but the effects are reversible but when it comes to sheep and goats, onions have absolutely no effect on them. There are certain breeds of dogs that are highly sensitive and cannot be able to handle the effects of onions and they are the Akitas and the Shibas.

**Figure 6.1.** A representation of garlic and how toxic it can be.

*Source: https://images.app.goo.gl/SkME4X4FbZZhzsFf6*

There is a compound that is known as N-propyl disulfide, which is an alkaloid that is found in onions, garlic as well as chives. This compound is mainly responsible for increasing the oxygen levels in hemoglobin as well as the membranes of the red blood cells and as a result, these two are targeted. When it comes to hemoglobin, once it has been injured during the oxidation process, it dissolves to form a precipitate of the Heinz body that can be observed in the red blood cells. In human beings, if a diet contains more than 25% constituents of onions then this can lead to the person becoming anemic.

Onions are usually plants that have bulbs and their leaves are usually narrow and linear as well as hollow and basal. When it comes to the stems of onions, they are usually not complex and they contain, not more than 3 bracts that have membranes. They have flowers that are of different colors and they include white, purple, pink as well as green and these flowers are usually parted into six perianths. The parts that make up an onion when crushed, they all smell like onions specifically, the wild onion that is scientifically known as *Allium candense*.

The toxins affect different parts of the body in different ways. The gastrointestinal effect include bad breath of any organism that consumes the onions, the urine as well as the milk also have an odor of onions. The musculoskeletal effects are weakness as well anemia which causes recumbence. To avoid the above effects it is important that the animals are left to relax and not stressed as well as they should avoid being given the onions until they are cured of anemia completely. When it comes to animals that have developed severe cases of anemia, then transfusion of blood needs to be considered.

Plants that are poisoned by onions have the following side effects in relation to their cardiovascular systems, increase in the pulse rate as well as the pulse might become faint also the animal is at risk of fainting due to lack of enough blood. In cases that are very severe, the animal can die especially due to lack of blood as well as the infestation of the animal by parasites.

The urine of a poisoned animal is usually dark as well as brownish in color a condition that is known as hemoglobinuria and it usually indicates the presence of toxins in the body. A lot of research has been conducted and it has proved that when cattle are fed with more than 25% of onions, they tend to develop anemia plus affected weight gain of the animals. Another example is the pregnant ewes who despite eating a diet that contains 100% onions are able to cope with the side effects of onions and they cannot die from poisoning that is brought about by onions and anemia induced by onions. When it comes to the sheep and the goats, their meat is not affected in any way in terms of quality and flavor even if onions are given to these onions for a long period of time.

*Asparagus:* The scientific name for this plant is *Asparagus officinalis* and it is a type of vegetable specifically the perennial type. It is mainly grown in the United States specifically in the department of plants that are termed to be agricultural between hardiness zone 3 as well as zone 8. This is where the temperature that is considered to be minimal is always above 40 degrees Fahrenheit. The part of this plant that is edible is the spear and it usually grows very fast. It can grow to a height of 5 to 6 feet and it can contain branches as well as berries. This plant grows in phases and they can be poisonous to animals as well as human beings if they are not handled and consumed with a lot of care.

When the shoots are young in the asparagus, they can be poisonous to a human being if they are eaten when they are not cooked. Most of the time they do not have very bad side effects and this research has been done by North Carolina State University Cooperative Extension Service. If and when the young shoot is cooked, the poisonous compound is eliminated and the plant becomes fit for human consumption. The stems of this plant usually grow up to 10 millimeters in terms of the diameter and these stems have proven to be toxic as well but not extremely toxic and they can make one become uncomfortable after they have consumed the stem of this plant.

There are some people who are actually allergic to this plant and once they have ingested this plant, their symptoms tend to be very severe as compared to other people who do not have this allergy. Allergies can also occur when a person touches the plant and when an allergic person comes into contact with this plant then they tend to develop a rash where the plant touched the skin. There is an inhibitor that is responsible for these effects and it usually contains sulfur which does not support growth and it is known as 1, 2, 3-Trithiane-5-carboxylic acid and this research and study was done by the New Zealand Society of Dermatologists. It has also been established that when a person consumes a lot of berries that come from this plant then they may experience side effects such as pain in the abdomen as well as cases of vomiting that may be observed.

The Asparagus fern which is very similar to asparagus only that this fern does not have a stem that can be eaten when it is still growing. This plant also produces berries that when taken in large amounts can lead to the same effects as the ones that are caused by the asparagus plant that can be eaten. When one comes into contact with the sap that is from this plant, a rash may appear but it is usually not a big deal because it clears on its own without any medical intervention. This plant is highly poisonous to both cats and dogs, this is because when the sap comes into contact with the skin it causes irritation that is not mild and when a person ingests the berries then it causes issues that are gastrointestinal. These plants are usually planted to be a source of food and they are harvested way before they can mature when it comes to the mature plants they produce some berries which are highly poisonous because they contain the following toxic compounds such as furostanol, as well as, siprostanol saponins.

**Figure 6.2.** Asparagus as a toxic plant.

*Source: https://images.app.goo.gl/RFhsYtpe4V6jJRN66*

## 6.3 POISONOUS WILD PLANTS IN THE UNITED KINGDOM

In the United Kingdom, specifically in North Yorkshire, these poisonous plants are displayed in a show which is known as The Handle with Care display and it is usually in the Harrogate Autumn Flower Show. This show allows people to view these deadly plants while maintaining safety. Statistically speaking, in the years 2014 and 2015 alone, the number of people who were admitted medically in England as a result of accidents that were related to plants as well as fungi was greater than 700. The person who directs this show claimed that these poisonous plants grow very quickly and they can easily find their way into gardens that are meant for vegetables and as a result mix up might occur which might lead to poisoning (Nyerges, 2016).

He also emphasized on the importance of people learning about these plants so that they can know which of the plants is dangerous as well as know how to deal with them in the right manner. The wild plants that are regarded to be highly toxic in the United Kingdom are:

### 6.3.1 Deadly Nightshade

It is scientifically known as *Atropa belladonna* which belongs to a family known as Solanaceae. The first people to grow this plant were the Europeans, people of North Africa as well as people of Western Asia. In the United Kingdom, this plant is usually known because of its ability to cause hallucinations as well as the toxic properties. This plant has foliage which is the leaves plus berries. These two parts are the ones that are usually very toxic and the compounds that make them poisonous are atropine, scopolamine as well as hyoscyamine. These berries that are produced by this plant are known to be very sweet as well as it has a very pleasant taste and as a result, a lot of people end up eating the berries without knowing the risks accompanied by them. These berries are highly poisonous. The medicines that can be used to reverse these effects are physostigmine or pilocarpine.

When this plant is ingested through the mouth, it brings about a lot of side effects such as, a woman who is pregnant can have complications with the pregnancy, and issues related to cardiovascular organs of the body as well as some disorders such as gastrointestinal or psychiatric ones. These plants have a very nice appearance and they are also said to very sweet and as a result, a lot of kids can easily be lured into eating them. The root, however, is the part of the plant that is highly poisonous but it is not always the case in different species.

Some of the symptoms that are observed when a person has been poisoned by the deadly nightshade are: the pupils of the person become very dilated, one becomes very sensitive especially when it comes to the light, one is not able to see clearly, it is hard for the victim to remain balanced as a result they may stagger, there is also a rash that develops, the throat and mouth of the individual becomes very dry, inability to urinate, inability to pass stool, the person might become very confused, he or she might see or hear things that are not there and there have been cases of convulsions.

When it comes to animals, this plant is capable of paralyzing the animal. Some of the animals do not experience any side effects like cows and rabbits. This plant contains a compound that is known to be anticholinergic which leads to memory loss while a person might not be able to learn.

Despite the fact that this plant is poisonous it can also be useful. It can be used in:

*Cosmetics:* The name of this plant comes from women who were believed to have named it, according to the Indians this name meant a lady who is beautiful. This plant was used to make eye drops that were used by the women to enlarge their pupils because this was considered to be a form of seduction as well as beauty. These eye drops facilitated the enlargement of the pupils by allowing the receptors that allow dilation of the pupil to work effectively by blocking those ones that reduce the size of the pupils. These days, the eye drops are not used because the side effects that are accompanied by the use of these drops are very terrible as they can lead to blindness as well as increased heart rates.

*Dietary supplements:* This use is very common especially in the United States and it means that this plant can be used to provide the body with some of the nutrients that are needed by the body very urgently. It is used to treat the common cold as well as flu among the people and they can easily be bought from any pharmacy and without a doctor's prescription. The use of this plant to make medicine for cold is allowed by the FDA. In as much as this plant is used a lot, it can cause poisoning because it contains a lot of contaminants.

*Medicinal uses:* According to the research that has been made, the use of this plant in treating various illnesses has not yet been conclusive but this plant has been said to contain atropine as part of its components and this was established in the 1830s where this component was acquired from the plant after the plant was purified.

Combination of wild belladonna alkaloids with phenobarbital in the correct way and the correct proportions helps the body to fight some allergies as well as it is used as a sedative. The FDA have established that, this plant can be able to cure the irritable bowel syndrome because of the component that is known as atropine. In as much as this plant can be used to make other medicines, it can cause harm to the person taking it, take a woman who is expectant as an example if she takes this medicine then she is at a risk of harming the baby. This medicine can also make a person to be constipated as well as the person is not able to see clearly.

*Alternative medicine and toxicity:* This plant has been used to make herbal medicines for a long time now, the medicines that are made from this plant act as pain relievers, this medicine has the ability to relax the muscles as well as reduce the inflammations that may occur in different parts of the body, it can also be used to treat issues that are brought about by periods, it can treat as well as it can be used to manage peptic ulcers, it can be able to counteract the effects of allergies as well as it can be used to prevent a person from feeling sick when they are in motion. It has also been used in the past as a treatment for upset stomach but still, research is being conducted in order to establish if all of this is true.

Between the years 2010 and 2016 in the United States of America, the people who take this medicine are advised not to use this medicine for children this is because the medicine could be poisonous and the effect could be that it causes seizures, a person has a hard time while breathing, one might sleep for a long time more than normal, the muscles of the person might become weak, they may become constipated as well as they may have a hard time while urinating or they may be very agitated for no apparent reason.

*Recreational medicine:* There is a plant that is usually related to this medicinal plant that is known as the jimson weed and sometimes it has been used for entertainment purposes, they poses the ability of making a person feel very relaxed as well as it can make a person hallucinate and that is why it is used as a recreational drug. A lot of people who have used this medicine have claimed that the hallucinations are not nice at all and the fact that it is used for recreational purposes can be very bad because it is easy for one to overdose and this might lead to death. The components that cause these effects are known as the alkaloids scopolamine as well as hyoscyamine. This drug has also shown that is can cause a lot of problems especially when it comes to memory because it can cause the disruption of the memory and as a result, the person might become very confused. The effects that are brought about by this medicine have been established to be only felt for a period of not more than four hours, while the hallucinations can be experienced for a period of up to four days as well the adverse side effects might be felt for a longer period of time.

**Figure 6.3.** Deadly nightshade as a toxic plant.

*Source: https://images.app.goo.gl/8uCWEfD7MEGLtYgG8*

### 6.3.2 Giant Hogweed

This plant is scientifically known as *Heracleum mantegazzianum.* This plant was initially grown by the people who lived on the west side of the Caucasus, which is found in Eurasia. According to the research that was made, this plant was brought to the British people around the $19^{th}$ Century but it was used as jewelry by the people. As the years passed by, this plant became very common among the people and it grew along the riverbanks. This plant has a certain compound known as furocoumarins that affect the skin in that the skin is no longer able to protect itself from the ultra violet rays of the sun which can cause skin cancer due to the high radiations.

When a person comes into contact with this plant and then he or she basks in the sun, skin rashes that are very serious can occur as well as a person may get severe sunburns as well as blisters but all in all, this only happens 24 hours after the person has been exposed to the plant.

This plant has a life cycle which has four phases' or rather four stages and they include:

- The pre-flowering plants; this occurs during the first years of the plant and it is accompanied by the seeds forming the leaves. This phase lasts for more years and the leaves keep on sprouting from both the roots and the seeds of the plant.
- The second phase is known as the flowering plant also known as the midsummer. This phase is characterized by the growth of a plant until it reproduces (flowering)
- The third phase is the seeds also known as the late summer as well as the early autumn. This is where the plant produces more than 20000 seeds.
- The last phase is known as the dead stems also known as the late autumn as well as it is known as winter. This is where the plants die after forming the seeds as a result only stems that are dry are left as well as the heads of the seeds are left standing.

When the plant is still growing especially during the first years, some parts of the pre-flowering plant end up dying especially, during the cold seasons like the winter and some of these parts are the leaves as well as the stems. This whole process is known as herbaceous perennial.

The estimated time that is taken by the plants to reproduce is at most 5 years and when the conditions do not favor the growth of these plants then the whole process might go on for a period of at most 8 years. In the Republic of Czech, it takes a period of 12 years for a plant to be able to reproduce and it is usually between the months of June as well as July.

According to the research made the month that is most likely to bring about the production of seeds is August. Statistically speaking a plant can be able to come up with 20000 seeds averagely and the ranges of how many seeds can be produced by this plant are approximately ten thousand to fifty thousand seeds by every plant.

Seeds are usually brought about by pollination specifically the cross-pollination and it occurs in more than one plant but according to the researches made, there is also the possibility of self-pollination. This type of pollination which is known as self-pollination is very important because the seeds that are produced from this process are usually very healthy.

The effects of coming in contact with the sap of this plant are very quick to arise and can occur after 15 minutes after a person came in to contact with the plant and it worsens after a period of 30 minutes and the effects can be seen for a period of time. Those who are in charge should always ensure that the people know the risks that are accompanied by the consumption of this plant and they should avoid this plant at all costs. When a person is dealing with this plant at a personal level then he or she is advised to ensure that they are completely covered to avoid any contact with the plant. If by mistake one comes into contact with this plant then that specific part that has been infected should be washed with a lot of cold water as well as soap, also they should ensure that they do not come in contact with the sun for a period of more than 48 hours.

**Figure 6.4.** A representation of Giant hogweed as a toxic plant.

*Source: https://images.app.goo.gl/SNYxz18h8RgdFzq27*

### 6.3.3 Monkshood

This plant is scientifically referred to as *Aconitum napellus*. This whole plant when ingested can result in death because it is very poisonous. It is advisable that if a person is to handle this plant, then he or she should wear clothes that protect them especially the hands so one can wear gloves. The toxic compounds of this plant are known as aconite as well as aconitine. When one handles this plant with bare hands, then the toxic compounds might remain on the hand especially, when a person does not wash his or her hands. This can lead to the person consuming the plant unknowingly and as a result, it might cause problems that are related to the stomach and finally, when the heart is affected it can lead to death. Even in small amounts, this plant is highly poisonous.

This plant can be grown in gardens because it is very attractive in appearance. It has flowers that are blue in color as well as it has inflorescences that resemble spikes. This plant has certain uses that include, it can be used to cut materials as well as it can be used to dry some materials. When it comes to the propagation of this plant, it is done using the seeds of the plant and it can also be done by the removal of offsets that are usually obtained from the rootstocks of the plant and this is done annually.

This plant in as much as it has a lot of benefits it has proven to be very toxic because it contains compounds that are fatal when one ingests them but in the past, this plant was applied on the arrows, as well as the spears, in different communities when they were out looking for food as well as when they were fighting their enemies. This was done by obtaining the sap from the plant and dipping the arrows in the sap and then shooting the arrow to the target. This practice was done in various countries like Italy and it was specifically meant for killing the wild goats. In the past , this plant was used to kill thieves, as well as the enemies, of the specific people but it was later proven to be a very crude method and as a result, anyone that was caught using this plant to kill anyone or growing this plant faced a death penalty.

This plant contains some very toxic compounds such as aconitine, mesaconitine, hypaconitine ,as well as jesaconitine, that can cause harm to a person if they use it. The symptoms of poisoning by this plant, occur very quickly and at times they are very instant and when one has used a large amount of it has been concluded that death is instant. If the ingested amount is not that large then without treatment it takes at most six hours for a person to succumb to the poison.

The symptoms that are accompanied with this poison include, the first signs are always, a person might feel very nauseated and they vomit and show diarrhea. After these symptoms have been observed other symptoms start manifesting and they include, a person might experience a burning sensation as well as a tingling sensation an also one might become numb in some areas such as the mouth as well as the face also the burning sensation might be experienced in the abdomen.

In cases that involve poisoning that is very serious then he or she is not able to move any part of the body and also all the limbs in the body seem to become very numb and weak. There are also other symptoms that include low blood pressure, the sinus becomes blocked as well as a person experiences some irregular heartbeats and the heart might stop. The other common signs and symptoms may include, a person may start sweating a lot, a person may start feeling faint, they might also have a hard time breathing, headaches might also occur and a person might become very confused. A person is likely to die from this poisoning because it affects the heart by making the heart beat in an irregular manner ,as well as the heart tends to stop beating as well, to make the respiratory systems and heart to not be functional.

When it comes to treating poisoning that is as a result of this plant, then the person needs to be closely observed especially when it comes to how fast or slow the blood is being pumped in the body as well the heart must be checked to ensure that it is beating the way it is supposed to. It is advisable to give the patient charcoal that has been activated and it has proven to work only if the person is given the activated charcoal within one hour after they have taken the poison. The most common antidote that has been tested and proven that it can work. It is specifically used to treat bradycardia.

Ingestion is not the only way that a person can become poisoned, when handling the leaves of this plant if a person does not wear protective clothing then the poisonous compound tends to be absorbed directly into the skin. In this case, the first symptom is a person might experience a tingling sensation at the point where they touched the leaves with. The poison slowly moves from the arms through the blood and soon it reaches the heart. A person might also become very numb and it can be treated with similar ways with when a person ingests the poison. Death has become a major issue because the organs in the body start shutting off and they fail to perform their functions.

**Figure 6.5.** A representation of the monkshood.

*Source: https://images.app.goo.gl/8SpPs4vTdQAg6mFs7*

## 6.3.4 Poison Hemlock

The scientific name for this plant is *Conium maculatum*. This plant has a lot of compounds that are highly poisonous and they are all alkaloids, they include, coniine, conhydrine, pseudoconhydrine, methyl-coniine as well as ethyl-piperidine. The consumption of this plant leads to vomiting that is very excessive and it is also known to make the nervous system become paralyzed. This plant has proven to be very fatal especially, when it affects the respiratory system.

This plant has proven to be very dangerous to anyone that uses it be it the animals or the animals like cattle, pigs, sheep, goats, donkeys, rabbits as well as horses. It contains compounds that are highly toxic and they are known they belong to the alkaloid groups. In as much as the leaves are very toxic, the seeds and the roots contain more poison. There are different ways that poisoning can occur and they include, a person might be poisoned because of ingesting the plant, when one inhales the fumes from this plant, as well as the poison, can be absorbed into the skin if a person touches the plant without any protective clothing.

When it comes to the animals, the leaves that they are given could contain this plant and so farmers are advised to be very careful in what they give to their animals. This plant is very tall and the stem is usually green in color and it has some purple patches and it does not have any hair. During the first year of maturity, this plant only produces leaves and after a period of two years, it yields flowers that are white in color and their appearance usually resembles that of an umbrella and it is often confused with the wild carrot plant.

The toxic alkaloid compounds that are found in this plant usually attack the central nervous system and this can be fatal to both human beings as well as animals. In animals such as calves, this poisonous plant affects the temperature regulation and as a result, it has proved to cause a drop in the temperature that is way below normal and it can result in death. In other animals, it has shown to increase the rate at which the heart is beating.

A lot of animals tend to ingest this plant because it usually has a very nice appearance. When it comes to human beings, this poison affects the kidneys and as a result, the kidney is not able to perform its functions which can lead to death. The effect on the kidney also prevents the passing out of urine and as a result, the urine is accumulated and retained.

After one has ingested these toxic alkaloids then the muscles of the respiratory organs start failing because the poison weakens the muscles. When it comes to this type of poisoning it begins slowly and slowly graduates in terms of paralysis. In cases that are very severe, the patient might be put on ventilators to help him or her breathe for a period of 2 to 3 days this is because this plant paralyzes the respiratory organs.

Even after the animals have been poisoned, they still go back to eat the plant. When a pregnant animal ingests the toxic plant then its young one will be born with a lot of complications, as well as it may be disabled, and the offspring cannot recover from the effects. The animals that are lucky can get treatment in the form of surgery but in most cases, they just die. There is no antidote that is available for this type of poison and therefore it is very important that people be careful and also the farmers should make sure that they exercise extreme caution. The farmers are advised to use herbicides that inhibit the growth of this toxic plant, as well as they, are advised to adopt the zero- grazing method when it comes to their animals in order for them to monitor what they are doing and what they are eating.

**Figure 6.6.** A representation of poisonous hemlock.

*Source: https://images.app.goo.gl/BDF4VaZpsN3fbryv5*

### 6.3.5 Foxglove

It is scientifically known as *Digitalis purpurea.* This plant is well known because it produces digitoxin, which is usually in form of glycoside in a medicine known as digitalis, this medicine is well known for its ability to stimulate the heart and this function was established in the year 1785. This plant is wholly poisonous and a lot of people know this. When a person ingests the leaves of this plant, the side effects are usually a pain in the abdomen as well as the mouth, feeling like one might throw up or rather one becomes nauseated, actual throwing up as well as it can make a person experience diarrhea. In the most serious situations, the following side effects have been seen the inability to see well ,as well as development, of issues related to the heart as well as the kidney.

The only time that a person can experience poisoning is if they take this plant in excess. When taken in excess it results in a number of side effects that include, nausea, a person might vomit as well as the person experiences a running stomach. Occasionally it has been observed that a person might develop jaundice as well as they may not be able to control their saliva, the rate at which the heart is beating might either be increased as well as it can be decreased, there is the possibility of fainting, a person might experience seizures and in most severe cases it can be fatal. This medicinal plant is known to reduce appetite and a lot of people end up misusing it especially, those people who are looking to shed some weight. In the past, it was used to treat diseases, as well as prevent them, but these days it has become very hard for people to still continue with this practice because the right preparation has not yet been established but over the years it has proved to be a very effective medicine as it can be used by epileptic patients to prevent their episodes as well as it has been used by various doctors who have patients that experience seizures and hence it has been established that this medicine has medicinal purposes.

The degree of toxicity solely depends on the type of species the plant belongs to. This plant has toxic compounds such as cardiac glycosides, as well as steroid glycosides, and as a result, this plant has been given various nicknames such as the witch's glove and the dead man's bells. When it comes to this plant, no part is safe for human consumption but in as much as deaths that are as a result of the ingestion of this plant have been observed, they are very few and rare. These plants are mostly ingested by children who are below the age of six years.

This plant does not affect the children like it affects the adults or adolescents. The first symptoms that are seen after a person has ingested this plant are pains in the abdomen, the person might vomit or experience the urge of vomiting, the person might hallucinate and the hallucinations are not pleasant, cases of headaches that are very severe have been observed and also the person might have a running stomach. In very severe cases, the following signs and symptoms are observed, their pulse rate tend to very slow as well as they are not regular, tremors have also been reported, it has also been established that the poison hinders good eyesight because a person cannot be able to identify different colors, cases of people convulsing have also been reported as well as the heart is also affected and the heartbeat rate is either increased or decreased and as a result, this can cause death.

This plant is often confused with a certain plant that is used to make tea known as the comfrey plant. Cases of deaths that are as a result of this poison are also common in children who usually drink water from sources as well as storage equipment that contain this plant. Even if the plant dries the poison is still left intact and any excessive ingestion can easily lead to death. It has shown to be toxic to both animals as well as animals.

The effects of this poisoning can be a person's heart can become blocked and also the rate at which the heart pumps blood decreases and it might as well lead to the increase of the rate at which the heart pumps blood. This is all dependent on the amount of toxins that a person or an animal ingests, as well as the condition, in which the heart of the affected person is in.

In as much as this plant is very toxic, it has some medicinal uses which include:

*Cardiac:* Digitalin is the term that is used to refer to groups of medicines that are obtained from foxglove. This plant has been successful in treating various diseases that are related to the heart and the specific component that performs this function is known as cardiac glycosides. This is all attributed to William Withering who was the person who came up with these findings in the year 1785 and it marked the beginning of modern methods of treating these diseases. This plant has been successful when it comes to ensuring that the rate at which the heart is pumping blood is under normal ranges and the heartbeats are regular, it has also proved to be able to increase the rate of the heart when need be.

This medicine is given to those people who have congestive heart failure where the heart becomes weak to a point that it cannot pump blood that is sufficient to maintain, as well as facilitate, normal body functions. As well it is used to treat the symptoms that are accompanied by this disease, a prescription is only given a medical practitioner specifically the doctors. The clinical trials were conducted and it was established by the Food and Drug Administration that digoxin is a medicine that is obtained from this plant and can be used in the treatment therapy of various heart conditions. This drug has also to be very effective in treating people who have a certain condition known as atrial fibrillation.

The College of Cardiology that is in America that is also known as the American Heart Association has supported the use of digoxin to treat patients who are experiencing heart failure that is as a result of chronic illnesses, as well as this drug, has proven to be very effective in increasing the systolic function and also ensuring that this function is preserved. This medicine has also been campaigned for use by the Heart Failure Society of America.

The use of this medicine when it comes to treating heart diseases has recently been opposed because it is not giving the desired results. It has also proven to be very lethal especially when it comes to females, the rate of death is very high and as a result, the use of this drug to treat various diseases has been highly questioned.

*Variations:* Some compounds that are present in this plant have been proven to be very efficient in the medicine sector. The compounds are known as digoxin and they can be used to make medicine that is later sold to people. Different pharmaceuticals have different names for the medicines that are obtained from these compounds and they go by the names Crystodigin and Lanoxin.

The medicines perform their functions by preventing the production of sodium-potassium ATPase. As a result, the sodium ions that are found in the body are able to increase in number and this results in a decrease in the sodium ions in the membrane of a person's cells.

## 6.3.6 Winter Cherry

Another poisonous plant that is found in this display is scientifically known as *Solanum Capsicastrum.* This plant also produces berries that are very toxic because they contain a toxic substance that is known as solanocapsine, which can be compared to the toxic alkaloids that are present in the nightshade plant. Small kids, as well as pets, are the ones who are prone to poisoning that is caused by this plant because the fruits of this plant are very attractive. Most of the time these fruits are not fatal upon consumption but they do have some terrible side effects and it is important for one to ensure that his or her kids do not come in contact with this plant. This plant requires a person to be very keen especially the adults.

## 6.3.7 Poison ivy

The scientific name of this plant is *Toxicodendron radicans* and it mainly causes allergies. This plant is native to the Asians as well as the people of Eastern and North America. This plant is well known for the fact that when a person comes in contact with this plant he or she becomes very itchy, their skin becomes very irritated as well as a person might develop a rash that is very painful.

The compound that is present in this plant that causes this problem is known as urushiol and it is usually a liquid that is very clear. This plant lives in different habitats, as well as different conditions and the animals, that mainly feed on it are birds that consume the seeds. This plant is commonly found in areas that are woody in nature but still allow the sun to pass through but only in limited amounts as well as it can be found in areas that have a lot of rocks, bare land as well as areas that do not favor the growth of any plant because of the geographical location as well as areas that do not have conducive climatic conditions for many plants to thrive in.

This plant can thrive in many soils with different pH such as acidic conditions, moderately alkaline soils. It can also be able to thrive in any climatic conditions but it does do very well in areas that have a lot of carbon dioxide in them because it is able to use that gas to reproduce as well as survive. The high concentration of Carbon dioxide allows for the plant to grow at a faster rate than normal and also it makes the plant more toxic and increases the reaction to the skin of human beings because of the toxic compound increases in amount.

This plant generally causes an allergic reaction to the person that has come in contact with it and in very severe cases it can lead to a more vigorous allergic reaction that spreads very fast and can be fatal, this condition is known as Anaphylaxis. Most of the people do not react to this plant but upon constant exposure, they might become highly reactive to the allergic compound found in the plant. When it comes to the rash that forms as a result of contact with this plant, it can last for a period of not more than twelve days but in conditions that are very extreme, it is possible for the rash to be seen for more than a month. According to statistics, it is approximated that more than 350000 people are usually poisoned by this plant in the United States of America.

This plant produces different compounds known as catechol and oleoresin which tend to be the cause of allergy. When this plant and compound come into contact with oxygen, it changes its color from a clear liquid to black in color. It usually sticks on the skin and as a result, a person feels very itchy and it slowly progresses until inflammation occurs that is red in color and later blisters might occur and it can be easily treated with a lotion that is medicated and is known as Calamine lotion as well as other solutions also some creams that help to relieve the itchiness as well as some soaps that help in making a person more comfortable. Doctors recommend certain products that help to relieve the itchiness like oatmeal as well as baking soda.

It should also be noted that the fluid that is released from this plant does not facilitate the spread of the poison because the fluid is from the body and not the plant. The poison can be spread from one person to another through contact or if the infected person touches a surface the poisonous compound remains there and is spread to the other person if he or she touches the contaminated surface. This is the reason why the poisoning can continue for a long time and the reason why a lot of people are affected even if they did not come in contact with this plant. When this plant is also burnt it can still because adverse allergic conditions because the fumes that a person inhales the rash will occur in the lungs and as a result the affected person is usually in a lot of pain and his or her respiratory organs begin shutting down and as a result the person might die. When a person ingests this plant, their digestive systems as well as their mouth, are affected and a rash develops within a period of one month after the person ingested the poison and this period is also dependent on how the rash is being treated as well as how severe the rash is.

Though very rare, there are a lot of situations where a person might require to be admitted to a hospital for further treatment as well as observation. If the plant dries up and dies it can still cause poisoning because the compound does not get inactivated and still remains active even after many years. If this plant comes into contact with things like clothes it can still cause poisoning if the person wears the cloth.

There are treatment methods that can be used in case of poisoning and they include, a person is expected to wash the affected area with a lot of water and soap. The water needs to be cold and also instead of using water and soap the person can use rubbing alcohol to stop the spread of the rash, as well as to prevent, the adverse reaction that results from contact with this plant.

Other medicated solutions may be used to provide comfort to the affected person.

**Figure 6.7.** A representation of poison ivy as a toxic plant.

*Source: https://images.app.goo.gl/aMEpavj1U5oynHodA*

### 6.3.8 Stinging Nettle

The scientific name of this plant is known as *Urtica dioica* and it belongs to a family of plants that are known as Urticaceae. It was first grown in Europe and then it spread to Asia to West and north Africa and now it is grown or rather it is found in every part of the world. In as much as this plant is poisonous in the past, it was used as a medicine that was used to treat various diseases , used as a source of food , in tea and used to make clothes in the past.

When a person comes into contact with this plant he or she becomes inflamed at the point of contact as well as the part that is affected experiences a burning like sensation as well as a stinging sensation that is usually not pleasant. There are several medicines that can be used to treat the burning sensation, as well as a stinging sensation, and these drugs are anti-itch medicines and they usually contain an antihistamine with hydrocortisone that are well-known for treating allergies as well as relieving itchiness in different parts of the body. Traditionally this problem is treated using another plant that is known as the dock leaves that usually thrive in the same condition that the stinging nettle does. They offer relief when it comes to itchiness as well as the burning and stinging sensation.

This plant has various uses and they include:

*Culinary use:* This plant can be used when cooking and it can be mixed with other vegetables such as spinach and cucumber because it has nutrients that are important to the body and they include vitamin A, vitamin C as well as it is rich in iron, potassium, manganese as well as calcium. A long time ago this plant was cultivated and harvested and used during the times when food was hard to find and it was harvested while it was still young.

The only way the plant can be used as food is when it is soaked in water to remove the chemicals that cause a stinging sensation without harming the person that is using the plant. This plant has been said to cause irritation in various parts of the body, even the urinary tract. The nutrient that makes this plant very important is protein because this plant contains up to 25% protein that is required by the body to maintain as well as ensure the proper functioning of the different organs in the body. When the leaves are dry they can be used to make a tea that has medicinal value and also this medicine can be prepared using the flowers of this plant. This plant can also be used in making various dishes such as soup as well as a puree. This practice is most common in North and East Europe.

It can also be incorporated into various doughs that are used to make different pastries.

*Competitive eating as well as a beverage:* This plant has been used in various countries as a competitive game especially in the United Kingdom where they have a specific day that is known as the world nettle eating championships where people compete and participate in the eating of this plant. In this competition, the competitors are asked to eat the plant when it is raw and as quick as they can, they achieve this by eating the leaves of the plant. The person that eats most of the leaves ains well as takes the shortest time possible to do this is declared the winner and is given a prize. This practice was introduced by two farmers who are trying to settle a dispute about whose farm had a lot of stinging nettles.

When the plant is still young it can be harvested and it can be used to make alcoholic beer.

*Traditional medicine:* This plant has a lot of medicinal value and in the past, it was used as a source of medicine by different people in different parts of the world. It was mainly used or rather it was given to breastfeeding mothers because it was believed to increase the production of the milk in women. This plant was also intentionally applied to the skin to cause inflammation as they believed that it was used to treat a condition known as rheumatism.

*Punishing criminals:* In the past, this plant was also used to punish those people who committed crimes and the criminal was beaten with a stinging nettle while they were naked and in front of people as they were being washed with water that was freezing cold.

*Making clothes:* In the past, this plant was used to make clothes because it contained fiber that was used to make different fabrics like linen. This practice is still being used today and statistically speaking it has been used to make a lot of clothes all over the world. During World War 1, the soldiers were made to wear clothes out of the plant because there was a shortage of clothes specifically there Germans what the people who used this plant as clothing.

This plant can also be used to dye different product because it produces yellow substances from the roots that are used to dye different items as well as it produces green like substances that are very effective especially when it comes to the dyeing process.

*Gardening:* As much as this plant is used as a vegetable it can be used as a method for attracting insects that are beneficial especially, in the growing of different plants in a garden. these plants are mainly grown in areas that are rich in the soil with a lot of nitrogen and this is an indication that the soil in the specific region can be used to grow other plants and the plants that are grown and able to thrive which makes work easier for the farmers. This plant can also be used to make fertilizers because it contains a lot of compounds that have nitrogen incorporated in them. They also contain other nutrients that can be used in making fertilizers such as magnesium, sulfur as well as iron. The fertilizer is usually low in phosphate but it still performs the required function effectively and efficiently and the fertilizer is used in gardens to grow other plants that are useful.

**Figure 6.8.** A representation of stinging nettle as a poisonous plant.

*Source: https://images.app.goo.gl/aMEpavj1U5oynHodA*

# 7

# *Domestication of Wild Food Plants*

## CONTENTS

## 7.1 INTRODUCTION

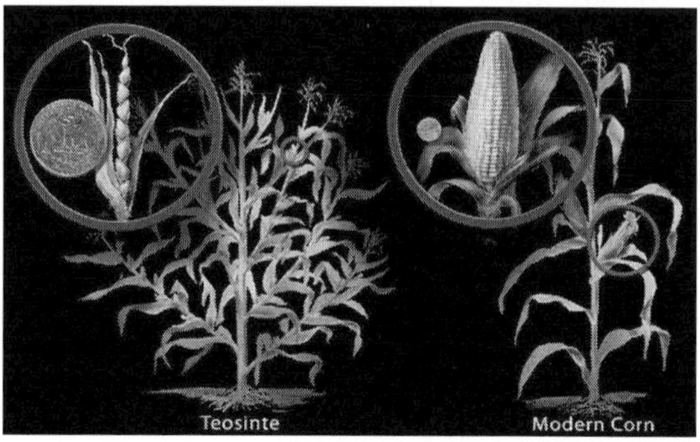


**Figure 7.1:** Modern corn was selectively bred from a distant wild ancestor.

*Source: https://theconversation.com/did-the-first-farmers-deliberately-domesticate-wild-plants-77434*

Plant domestication is an iconic technology introduced by people during the change from hunting and gathering lifestyle to agriculture (roughly 12,000 years ago). The early stage of domestication is a selection to increase adaptability to human cultivation, consumption and utility of food crops. Through multidisciplinary domestication studies, a wide range of topics is covered, including the geographical and ecological geneses of crops, the heritage of domestication traits, the spread of crops from their zones of origin, and the time and speed of domestication. With the ever-changing adaptation of human beings, plant domestication research has impacted immensely on the continuous improvement of crops (Daniel Zohary, Maria Hopf, & Ehud Weiss, 2012).

Plant domestication is an evolutionary process in which plant populations have become accustomed to human management and control. For researchers, domestication is usually regarded as the end terminus of a continuum which begins with the exploitation of wild crops and continues with cultivation from the selection of wild plants, but it is still genetically indistinguishable from wild plants, and finally its adaption to agriculture. Ecological studies are conducted through conscious and unconscious people's morphological choices to distinguish domesticated species from their wild ancestors through genetic differences.

According to the native communities, collecting plants from wild sources to grow on farms (fields and home gardens) is commonplace and is an activity continuously carried out under various agricultural ecosystems. Through this process, many plant varieties, local species and cultivars have been developed to meet human needs for food, fiber, pharmaceuticals, building materials, and so on, or research for a species or group of species such as yam, barley, rice, baobab, acacia, Fonio and leafy vegetables (Daniel Zohary, Maria Hopf, & Ehud Weiss, 2012).

## 7.2 PLANT DOMESTICATION CONTEXT

The evolutionary lineage that led to modern humans (*Homo sapiens*) has various key inventions, like the development of progressively sophisticated tools, mastery of fire, mastery of writing, and the present digital revolution. These inventions include agriculture or the changeover from hunting and gathering (HG) to crop cultivation. This change has triggered a change that can't be overestimated in various areas of life on Earth, such as biological and physical environments, and obviously the human lifestyle.

Therefore, it is not astonishing that this transition is known as the Neolithic Revolution. There are many signs of change initiated by the widespread introduction of agriculture. For instance, many of the famous "wild" landforms on earth are really agricultural landscapes: such as the rice terraces in Yunnan, China, rolling cypress-covered land in Tuscany, Italy, and the bright, high-altitude plains of Palouse, northwestern United States (Harlan, 2011).

Now, 38% of the entire land area is used for agriculture, which makes it the most significant land use at present. For instance, forests (most forests are subject to some kind of human development and utilization) account for only 30 percent of the land expanse (FAO, Land Use Database). Changes that occurred more or less simultaneously with the origin of plant domestication include: sedentary (as opposed to nomadism), the emergence of villages and eventually cities, ceramics introduction, and the transformation from a more equal society to a hierarchical one. The domestication of plants means more food and therefore, a larger population. The resultant food surplus is the premise of a settled, politically consolidated, socially stratified, economically diverse, and technologically innovative society (Harlan, 2011).

## 7.3 THE MODEL OF DOMESTICATION AND DOMESTICATION SYNDROME

Essentially, domestication is an evolutionary procedure of choice to adapt to changing environmental conditions, because are not just consumed but cultivated as well. Cultivation is an essential condition for plant domestication to happen. The selectivity on plant populaces in the wild environment is largely different from the subsequent cultivation modern environment. In the wild environment, there is a strong competition between plants for light, moisture and soil nutrients. Contrarily, agroecosystems are categorized by lesser competition, even in renowned, diverse arrangements (such as agriculture, forestry, and gardens).

This lessened competition is then transformed into greater productivity of harvested tissues (example, cereals, leaves, roots) through the improved distribution of photosynthetic products and a greater harvest index (the share of harvested biomass to overall biomass). In terms of cultivated land and unit biomass per unit time, more efficient production systems need to be developed. The overall objective is to grow the biomass harvested, even though the overall effort needed to harvest the biomass is reduced. Therefore, several features have been chosen to achieve this objective (Henry, 2010).

These characteristics include: (1) reduction of seed dormancy, resulting in a more uniform stand during sowing; (2) more compact growth habits, fewer branches, shorter branches, which lessens intra-species competition and makes harvesting more synchronized; (3 ) Propagation has a tendency from heterogeneous reproduction to inbreeding to asexual reproduction, which makes the reproduction less reliant on abiotic and biological factors; (4) minimal seed breakage, thereby limiting grain loss during harvest. Third, whether it's used as food, feed or fiber, humans not only serve as farmers, but also as users of their production, no matter the case they all play an important selective role.

One key feature here is the attractiveness of humans to novel things, such as the color or shape of new seeds and fruits, which leads to an amazing morphological diversity of certain crops, such as the developmental changes of cabbage (Brassica: Red Cabbage, Brussels sprouts, and cauliflower); size, seed color, and shape in some legumes; and fruit size, color, shape, and texture in the Rosaceae plant group (example, apples, pears, cherries and plums).

Generally, domestication leads to an increase in the overall size of harvested organs, whether these are fruits, roots, leaves or grains. Sometimes the increase in size may be quite significant since the variance between wild and domesticated plant types maybe 10 to 20 times. Plants often contain toxic substances that can perform multiple functions, such as protection against illnesses or pests. In times of domestication, the compound concentration in all or some partially domesticated gene pools is usually reduced, for instance, as can be seen by the lessening of cyanoglucoside in cassava and lima beans (Daniel Zohary, Maria Hopf, & Ehud Weiss, 2012).

The presence of these characteristics with related traits is the primary feature of domesticated plants, and shows a certain level of consistency between different or even unrelated crops, which makes scientists call them "domestic syndrome". However, sometimes there are significant differences between crops, based on their general history and reproductive methods. Perennials, like nut trees and vines, usually show a more restricted difference between wild and domesticate types due to their prolonged juvenile period, long life phase, high gene flow, and prospective for asexual reproduction. For instance, the key domestication traits of grapes are their androgynous properties (different from male and female reproduction).

This change is managed from a single location. Therefore, in general, the domestication pattern of perennial crops is composed of fewer disparities between wild ancestors and domesticated offspring. The expression of fundamental features is also typically restricted. For any given crop, there could be multiple domestication syndromes, based on local production conditions. For instance, a different domestication pattern can be imagined in the form of a raw product for grain biofuels (which is presently the case) or its nutrients. The latter may involve more branched plant types, thereby increasing the yield of stems or leaves, instead of grains (Daniel Zohary, Maria Hopf, & Ehud Weiss, 2012).

## 7.4 WHERE DID PLANT DOMESTICATION HAPPEN?

As an invention, the domestication of plants is entrenched in the heritage of agriculture itself. Traditionally, the main areas of knowledge around the onset of agriculture are biology, linguistics, archaeology and history suggested by researchers. Ever since a wide range of research fields have been implemented to solve the mystery of the origin of agriculture. Archaeological botany attempts to describe the crops from Neolithic sites and the archaeological remnants of their wild ancestors. Of specific interest is the changeover from the cultivation of wild vegetation (pre-domesticated cultivation), to the cultivation of real domesticated items in archaeological records.

Some basic traits have been recognized, mainly involving the breakage of grains, like wheat and barley. Hereditary evidence is mainly based on studies of contemporary wild and domesticated plant populations. An important factor is to control the distinctiveness of wild ancestors of each crop. First, through artificial hybridization, the survivability and fertility of the offspring between the crop and the assumed wild ancestry is determined, which has been achieved for most crops.

Since domestication only occurred up to 10,000 to 12,000 years ago, the differentiation between crops and their wild ancestors is limited, and it is expected that in most cases their hybrids will have regular viability and fertility, but with exception of polyploid sequences, for example as wheat. For latter crops, although the ancestors and offspring may have been closely related, hybridization comprising of different ploidy levels typically doesn't produce viable or fertile offspring.

Then, other analyses comprising molecular markers may determine the identity of crop ancestors. All molecular pointers are eventually based on deoxyribonucleic acid (DNA) sequences, whether these are proteins (such as enzymes (synthase) and seed storage proteins) or the DNA sequence itself. Due to advances in DNA sequencing technology, more and more plant populations are characterized by direct sequencing (genotyping by sequencing). Such analyses are similar to paternity testing, revealing the smallest amount of domestications, plus in some cases, presumed acclimatization areas with a higher degree of acclimatization across the entire distribution area in the form of wild ancestors. Recent illustrations of this ability comprise of the identity of putative domestication zones for plants in the Mesoamerican epicenter of agricultural origins, like maize, pepper, and common bean in various zones of the Mexican terrain (María de Cortes Sánchez-Mata, & Javier Tardío, 2016).

According to scientific analysis, the origin of domestication of wild crops has been determined as Central America, Southwest Asia and ancient China. For example, corn and kidney beans were domesticated in various watersheds found in Western Mexico, however, they were eventually integrated into the Milpa ("Three Sisters") planting system through pumpkins after domestication. Durum, Einkorn, and bread wheat are domesticated in various areas of or near the Fertile Crescent, however, they also grow in parts of Tibet and Central Asia. In China, there might be multiple domestication zones, one of which is around the Yellow River Basin, common millet. Also, foxtail millet and Asian rice grown around the Yangtze River Basin are some common examples of plant domestication.

## 7.5 FACTORS THAT CONTRIBUTED TO DOMESTICATION OF WILD PLANTS

The dry season might have been the driving force for the cultivation of wild plants due to the improved need to develop food storage, whether they are cereals or roots. Domestication is not limited to the Mediterranean and tropical dry steppes and forest biomes. Different other environments also contribute to crops, such as the plateau environment (for example, coffee on the Ethiopian highlands; plus various root crops such as the Andean potato). Coastal regions also contributed to the development of coconut, beets, cotton. While some temperate forests might have contributed to fruits and nuts, like apples, cherries, pears, and walnuts (María de Cortes Sánchez-Mata, & Javier Tardío, 2016).

The tropical rainforest is most likely the turf of sugar cane, bananas and plantains, mangoes and cocoa. Domestication centers often occur in regions with high biodiversity and biodiversity hotspots. Therefore, actual or potential farmers seek opportunities by domesticating potential crops from local biodiversity, no matter where they are found. In this concern, it might not be startling that some species or collections of related plant species have become domesticated many times. For instance, five types of kidney beans have been domesticated in Central America and the Andes, of which common beans as well as lima beans have been domesiticated twice.

Three other domesticated kidney bean Species-Runner Bean, Year Beans and Tepary Beans were domesticated in different ecological environments, reflecting the possible choice of farmers to choose different domesticated products. Other common models are Capsicum varieties (five species tamed in different regions of Mexico, South America and Central America, and; cotton (four species tamed, two species in the old world, two species in the new world, as well as wheat (where three species have been domesticated within the Fertile Crescent (María de Cortes Sánchez-Mata, & Javier Tardío, 2016).

## 7.6 MODERN DAY DOMESTICATION OF WILD PLANTS

It's projected that there are approximately 300,000 to 500,000 higher plant species, of which approximately 368,000 have been classified or described. Still, many species are not yet known to science, and perhaps one- third of them are even in danger of extinction. The amount of plant species utilized for food in pre-agrarian human society is just a small part of the domesticated plant kingdom.

The procedure of crop domestication is based on human cultivation practices and selection driven by the agricultural environment. About 2,500 species have been domesticated to a certain degree, while 250 species are believed to be entirely domesticated. People depend on a small number of crops, like rice, wheat, soybeans, corn, and potatoes, which comprise the majority of people's dietary intake. A total of about 10 to 50 plants provide 95% of the globe's caloric intake. While most food is focused on just a few plant species, this is still a key factor for combating the vulnerability of the globe's food supply to climate change and possible plant disease outbreaks (Henry, 2010).

Nevertheless, crop wild relatives or (CWRs) are still the largest genetically diverse resources, can be used for crop improvement, and have been used for major genetic illnesses, pest-resistance and abiotic stress forbearance. Nevertheless, a large number of plant species with good traits have not been domesticated. While gradually understanding the genome and biological context of the domestication process, it's possible to apply more effective selections for the domestication of wild species.

Because many wild taxa have been locally adapted to specific habitats and have important genetic diversity, this may lead to the production of new crops, while increasing the possible adaptation to new production environments and helping humanity when facing climate change that is affecting the growth of some traditional plants (Henry, 2010).

Although conventional breeding through cross-breeding and progeny selection, coupled with marker-assisted selection, maybe the most viable method, the latest advances in genome editing techniques, genome selection, and plant inoculation of beneficial microorganisms may help hasten this process. Come the future, this can be achieved through the transformation of key domesticated plant genes from wild to domestic forms.

- Domestication of plant types from scratch. This includes not just working with wild varieties of existing crops, but also domesticating completely new plant species with appropriate traits. This work has led to the evaluation of such crop species and traits appropriate for targeting. Knowledge of biology and ethnobotany can help determine suitable candidates. The process will provide prospects to adapt to local crops better.
- Application of crop wild relatives for expanding the genetic variety of crops is another area of concern. As it's widely known, wild relatives of crops have greater genetic assortment compared to crops, which may effectively make existing crops more resistant against biological and abiotic factors, like climate change.
- Perennial plant species growth. The process of domestication in the past has come at the cost of growing habits every year, at the expense of challenging agricultural practices, plus eventually has a negative influence on the environment. The addition of perennial crops (such as cereals) make the environment a bit friendlier and viable.
- New domestication syndrome and reimagined models. During the domestication procedure, the common feature set (domestication syndrome) becomes modified. Nevertheless, this inevitably reduces the genetic and biological scope of today's crops. Modern agriculture meanwhile has diverse requirements for planting and harvesting than previously. In addition, scientific understanding and breeding trends have recognized new crop phenotypes.
- The relatives of weed crops comprise a special example of CWR, which not only provides opportunities for breeders to use but also poses problems for farmers. With regard to crops, closeness to relatives of weeds can allow gene flow, transgene escape, plus contamination of crops and seeds. However, locally adapted weed species may provide the source of the genes needed to confer local adaptation, and can even be applicants for 'de novo' domestication of new ideologies.
- In polyploid crops, the progress of genetic knowledge, as well as genomic resources (such as complete genome assembly), is especially challenging, pooling attention to wild, diploid relative plants as a possibly attractive model system. The genetic context of the domesticated trait variant may be easier to analyze at the

diploid stage. The key domesticated traits can originate from particular diploid ancestors and thus exist in particular sub-genomes.

## 7.7 GENETIC SELECTION FOR WILD FOOD PLANTS

**Figure 7.2:** Genetic wild food plant modification makes them more resilient and suitable for human domestication.

*Source: https://allianceforscience.cornell.edu/blog/2018/03/food-genetically-modified-way-draw-line/*

In the opening sections of "The Origin of Species," Charles Darwin presented the concept of wild selection, similar to domestication. The significance of domestication to Darwin's thoughts is even apparent in his early sketches. Darwinians claim that examples of domestication are crucial to the formation of his theory. However, for Darwin, domestication is not just a useful analogy: he believes that domestication is an adaptive model from which inferences around the nature of diversity and assortment in wild systems may be drawn.

Darwin's proclamation of the significance of domestication in the comprehension of the evolutionary procedure wasn't universally accepted, however. Among the most vocal detractors of Darwin's opinions on plant domestication was Russell Wallace, Alfred. Even in their combined publication mentioning the model of wild selection, Russell denies the significance of domestication. It's seen, then, that no implications as to variations in a form of nature may be inferred from the opinion of those happening among domestic plants. It's continually been considered a drawback in Darwin's work, which he based his model, mostly, on the indication of variation in tamed plants (Harlan, 2011).

Wallace found that two features of domestication were wrong because it was an inspiration for comprehension adaptation in nature. He first argued that this analogy is flawed: manual selection needs an intelligent selector, and there is no such force in wild systems. In addition, he insists that the choice itself is essentially unique, resulting in fundamentally unique mutations. He mentioned that domesticated plant species are irregular, abnormal and artificial; they're subject to variations that never happen and never happen in a wild condition: their existence relies fully on human care.

Both of Wallace's arguments can find modern listeners, from people who see that humans consciously choose to have a fundamental difference between human and wild systems to people that trust the variation of domesticated species is different from nature. However, Darwin directly mentioned these ideas with a logical understanding of unconscious decisions. He divided human-mediated decisions into two sections: structured choices consisting of "systematic efforts for modification of varieties based on pre-determined st\andards" as well as unconscious choices" wild selection retains the most valuable characteristics and disregards the less valued attributes.

He believes that unconscious selection is not very different from plant wild selection. People have changed the conditions for the survival and reproduction of cultivated species exist. Even if the cultivator has no choice or predetermined goal, this change will exert a choice on the population. While the term "unconscious choice" has not been used for quite some years, crop evolution researchers still recognize its basic role in domestication. Researchers believe that the initial stage of domestication is entirely determined by unconscious choices. Modern workers have widely cited the core role of unconscious choices in achieving the observed variation for domesticated species, sometimes called automatic or unconscious (Harlan, 2011).

The phenotypic changes related to adaptation under domestication conditions are enormous. Most of these variations are shared across different domesticated crops. This change is regularly known as "domestic syndrome". The common characteristics of domestication syndrome are bigger fruits or grains, reduced branches, gigantism, loss or decrease of seed dispersal, loss of seed dormancy, variations in photoperiod sensitivity, as well as loss or lessening of toxic compounds. For example, a series of key cereal crops (rice, wheat, corn, and barley) brought about by domestication have experienced certain parallel phenotypic changes, including lessened seed dispersion, minimal branching and tillering, reduced seed dormancy, and synchronized seed maturation. The grain size increases and the inflorescence is also larger (Harlan, 2011).

### 7.7.1 Two Approaches to Finding Adaptive Genes

Dramatic changes in phenotypes related to domestication are important not just as examples of evolution. They have a broad economic and social impact. There is great interest in discovering genes and genome mechanisms that cause changes in domestication-based phenotypes since their identification can facilitate trait manipulation from improved breeding strategies.

To date, most studies have followed what's called the "top-down" approach, which starts with the phenotype of interest before passing through genetic inquiries (like quantitative trait loci (QTL) and linkage disequilibrium or (LD) mapping to identify disease-causing genomic regions. Another approach is to start from Darwin's domestication point of view, beginning with the model of adaptation, and then bottom-up. For this method, population genetic procedures are used to examine adaptive signals in a group of genes, while traditional molecular methods are used to transform from typical genes to phenotypes.

To date, all successes in identifying potential genes for adaptive variations during domestication have come from top-down methods, starting with phenotypes, then using genetic analysis to discover genomic parts, and finally finding candidates responsible for the phenotype gene. The most effective method to find these genes is QTL mapping, but the association method or LD method is highly favored in the plant genomics community. Although providing a comprehensive overview of QTL and LD mappings is beyond the range of this article, it mentions some empirical discoveries and highlights some challenges that bridge the space between genotype and phenotype.

*QTL Mapping:* Given a feature of interest, the QTL mapping is the first (and still the most commonly used) method that can be used to locate the genetic context of a plant trait. QTL mapping has caused fundamental achievements in the identity and cloning of genetic structures with wild characteristics. The most famous examples are tomatoes and corn. During the mid-1980s, scientists began to perform QTL analysis on the fruit quality of wild tomato and domestic tomato hybrids and identified six QTLs.

Through extensive mapping work, they could isolate an area containing the main QTL fruit weight 2.2 or (fw2.2).

They also established the phenotypic impact of fw2.2 through transgenic analysis. Around the same period, Doebley and colleagues mapped the differences in plant structure and plant yield between corns and its wild ancestor, teosinte. Successive mapping and mutation examination led to the division of the primary genes managing the phenotypic variation between corn and teosinte, including the genes teosinte branched1 or (tb1), and teosinte hull system (tga) that control lateral branching, which led to inflorescence structure difference.

These accomplishments highlight the worth of the QTL method, however, the technique is not without certain limitations. For instance, it may be hard to identify mapping numbers for perennial, inbreeding, and asexually reproduced crops. Therefore, it is somewhat difficult to study certain crops, like bananas or palm trees, through the QTL method. It's also key to note that the outcomes of QTL analysis typically rely on the environment, and the parental line applied in the hybridization. Therefore, extra care should be taken when interpreting the generality of QTL, especially in the case of manifold domestication or local variation.

There are many statistical problems, the most essential of which is the reduced ability to precisely estimate the quantity and scope of QTLs. The finding has been known as the Beavis influence. Although it has not been proven that this limitation is problematic for cloning plant genes with large phenotypic effects, statistical power has generated significant attention to more classical quantitative traits (such as size, weight, and yield). QTL determination of phenotypic effects and statistical issues become even more difficult to estimate complex factors like epistasis.

QTL researchers have provided and shall continue to provide substantial utility for identifying genetic factors and genome regions that contribute to the phenotype. In addition, as genomic data of more species is obtained, the identification speed of this gene will continue increasing. This increment was already perceptible in the 2006 publishing year and witnessed a surge in gene separation that led to major phenotype differences between domestic plants as well as their wild ancestors. While not only because of the QTL method, the genes separated in 2006 also comprised of two genes that impact rice nutrition, among them, is rice color gene, wheat breaking gene, as well as the aging gene.

Even so, just a few genes were separated by these methods, and due to a large amount of capital and human capital needed for QTL research, plus the socioeconomic significance of a comparatively small amount of crops, the overall output was surprisingly small. Additionally, the gene separated so far has a rather immense effect, specifically "low lying fruit". To recognize and clone non-effective genes may need increased effort.

*LD Mapping:* To overcome some of the drawbacks of QTL analysis, plant scientists have turned to LD mapping as another method to classify genomic regions that contribute to the phenotype. Practically, LD mapping may be divided into two kinds, each type focused on different levels of genetic analysis. Same as most QTL methods, the first method aims to identify whole-genome variations associated with phenotypic variation. It requires measuring the genetic variability of markers representing many genomes and testing the phenotype-genotype relation of each marker. The alternative form of association analysis tries to identify the causative genome mutation(s) which affect phenotype; many latter studies usually concentrate on variation in some candidate genes, instead of entire genomes.

The main benefit of LD mapping is the fact that it can depend on population samples. There's no reason for crosses or large amounts of offspring. This is a clear benefit for the research of bananas, palms and other durable multi-year biological plant species, and generally, it allows for the research to proceed much faster. Additionally, the population sample might contain more information than the traditional QTL mapped population (that is, all the information that happened in the evolutionary heritage of the sample). Consequently, the phenotype-related phenotype might be smaller compared to the chromosome sections in the QTL populace, which theoretically offers a greater localization resolution.

It is the same as the QTL method, but multiple functions of the experimental design must be carefully considered when performing LD mapping. For start, it's important to distinguish true relations from statistical noise, a big sample size would be required to enhance statistical power and rectify several tests. Even with big sample size, scientists may need to assume that the impacts of a given mutation are additive. Examining the epistatic interrelation between multiple markers further intensifies the issue of multi-testing. One method to mitigate the problem is testing the relation between haplotype and phenotype (and "haplotype module"), rather than a particular marker. However, unless haplotypes are inferred experimentally, such as in self taxa like barley or rice, the essential computational implication of haplotypes may prove an obstacle to this method.

Another design concern is the source of samples. Geographical structure and other deviations from panmixis may lead to false associations, where genotypes are related to geographic regions rather than phenotypes. This shall be particularly challenging for phenotypes that vary by geographic areas, like flowering time and photoperiod sensitivity. Most important crops (such as rice, soybeans and barley) come from wild populations having a wide geographic structure.

This structure is also often mirrored in domestication, particularly when multiple independent domestications are involved. However, for many crops, there's little information available on domestication or populace structures of wild populations. The noticeable exceptions are barley, rice, wheat and corn whose domestication history is becoming clearer. Research on human diseases has shown that basic studies on demographic histories and population structure is critical to the realization of LD (linkage disequilibrium) mapping for plants.

The ultimate design challenge to consider here is the density of tags (typically SNPs). When genotyping disease-causing mutations, LD mapping research is very powerful. In case the disease-causing mutation isn't genotyped, it may still be identified by the marker found in the LD with the disease-causing mutation. Nevertheless, the degree of LD may vary greatly between plant species, between genomic parts of the plant, and between population samples.

The circulation of LD is furthermore impacted by homologous genome transformation, which mostly damages the short-term LD pattern. Research design, statistical inquiry, and control of biological tests (such as population structure) are rather active research areas, but despite the lack of background information on the extent and geographic structure of LD in the population, some large-scale plant LD plotting studies are still underway.

The inherent challenges of LD mapping have been emphasized in the scientific literature. The researchers confirmed flowering times and pathogen resistance for various Arabidopsis QTLs, however, also mentioned a higher false- positive rate. Staff using wild-captured plant populations couldn't verify the correlations present in laboratory populations, signifying that no matter the sample size, genotype SNPs numbers, or research design, not all results are replicated. Additionally, it may be impossible to miss reports which failed to detect the relationship between a candidate genome and the final phenotype.

The problems associated with LD mapping are mirrored in research, scientists have confirmed flowering durations and pathogen resistance for various Arabidopsis QTLs, however also mentioned the higher false positive rates. Those using wild-captured groups of massive fruit flies were incapable of verifying the relations found in laboratory populaces, suggesting that some outcomes may not be replicated despite the sample size, genome SNP number, or research design. In addition, reports that failed to identify the relation between the candidate genome and the target phenotype may be missed.

Despite these shortcomings, LD mapping has achieved some success. An early example successfully linked the phenotypic variation of barley malt quality to the haplotype variation of the β-amylase double gene (a locus of starch hydrolysis). The difference in the coding area of barley β-amylase 2 affected the thermal stability of enzymes. SNP genotyping established that the varieties having high malt quality or high thermal stability contained the same haplotype.

Resequencing of candidate genomes has also been applied in rice and millet for determining the genetic context of waxy and sticky cereals. Genome variations at the waxy (grain-bound starch synthase) locus lead to changes in the amylose content of the endosperm, producing cosmids that are common in southern and eastern Asia. LD mapping furthermore has been used for verifying relations inferred from QTL and other methods. One prospective bearing for LD mapping is to use artificial populations derived from comparatively few creators, this helps LD and QTL mapping in a given population while reducing complexity because of population structure.

## 7.8 TYPES OF WILD FOOD PLANTS THAT HAVE BEEN DOMESTICATED

- *Potato beans:* Potato beans are one of the wild species. Also known as Apios, it's as versatile as potatoes, rich in protein like beans, has a subtle flavor like starch, and performs well in dry and moist soil. There are many other similar plants in the legume family. Apart from potato beans, there are about 18,000 legumes in the world. They are rich in protein and help to fertilize the soil. However, there are less than 50 domesticated foods, and usually, people only eat half of these foods. Scientists have assembled a list of potential candidates such as: marama beans, lupine, yehub nuts and different other so-called orphan plants and wild edible crops that can change the course of agriculture if somebody turned them into reliable plants.

- *Fireweed:* Fireweed, or scientifically known as *Chamerion angustifolium*, refers to an edible plant native to the northern hemisphere, sometimes referred to as Rosebay Willowherbin UK, and is also known as the great willowherb in some areas of Canada. It has upright red stems, as well as unique leaves, whose vein pattern is round and uninterrupted at the tip of the leaves.
- *Dandelion:* Dandelion, you may be thinking: "I can't eat this!" However, dandelion, or more scientifically known as "Taraxacum", is actually completely edible. It means its flowers, leaves or even roots are edible. Dandelion leaves are usually 5-25 cm long or much longer, their inflorescence heads are yellow or orange, opening at daytime and closing at night.
- *Chickweed:* This plant is scientifically called "Stellaria medium" and can be applied as a cool herbal and safe, nutritious food plant. Chickweed can be recognized by fine hairs found on only one flank of a banded stalk with fine hairs around its stem. Many believe that it makes one of the best salad compounds.
- *Curly Dock:* Rumex Crispus is often called "Curly Dock", if you are in a confined place and there are no other palatable options, then it's a safe snack option. Even though the plant's leaves can be said to be appetizing and the main source of vitamins A and B, it's best to eat in moderation because it can trigger urinary tract irritation when consumed in large amounts.
- *Asparagus:* Asparagus is easy to recognize, and its fleshy green spears are both sweet and delicate and can be consumed raw rather than cooked. It's a great source of potassium and vitamin C. It must further be noted that wild asparagus stalks are thinner compared to those found in grocery stalls. Moreover, the average harvest length for these crops is typically 6 to 8 inches.
- *Chicory:* The member of the dandelion plant family is bushy, usually with bright blue flowers. This plant's leaves can be consumed raw with flowers, however, the roots can't be consumed unless cooked.

## 7.9 THE FUTURE OF PLANT DOMESTICATION

Domestication has made mankind the first planet species to have a safe and reliable food supply, thereby promoting the advancement of culture, technology and medicine. All aspects of modern society are built around it. However, in the process, somehow humans stopped innovation. Even so, small but dedicated teams of researchers are nowadays quietly attempting to grow new plants. For instance, they are mixing up new varieties of wild sunflowers producing larger, oilier seeds that don't require much water. They are researching the genetic rewriting of chickpeas, choosing traits that can help humans thrive in the age of global warming (Henry, 2010).

Climate change makes this mission imperative; the genetic evolution is making this practical, it represents the next step towards the green revolution. Domestication basically is evolution under mankind's control. When nature becomes in charge, the model of evolution chooses plants based on attributes that are favorable for survival. People choose traits that promote yield, predictable growth, and flavor and disease resistance. Humans began doing this simply because they could. During the Paleolithic age, climate variations made it hard for people to over-depend on particular crops.

Domestication is evolution under human control. When nature gets in charge, the theory of evolution chooses plants based on features that are conducive to survival. We humans, choose traits that encourage yield, flavor, foreseeable growth and attack resistance. People from different cultural backgrounds have actively managed various wild plants from the earliest days (Henry, 2010).

From the thousands of plants (about 50,000 edible plants) that people typically used in prehistoric times, our ancestors selected only a small amount of all grass as the cornerstone of their diet. The early domesticated plants however have minimal similarity to their wild ancestors. For example, 10,000 years ago, in present-day Mexico, farmers planted a weed known as "teosinte" and formed corn. Teosinte's small ears have just a dozen kernels, meanwhile, there are about 800 heads of modern corn. Selective breeding changed a grass into a staple food with a high starch content that we know as corn today.

It looks simple, however, humans have not domesticated new staple foods for 1000s of years. Nevertheless, domesticated food crops look the same to each other. Most of the features we choose are the same, regardless of species. Currently, we want crops to keep seeds rather than falling to the earth, which plant researchers call "shattering". We wish these seeds to grow and germinate when they are sown, and we wish all seeds to germinate at around the same period. These characteristics together constitute what researchers call "domestication syndrome", which can distinguish corn from teosinte.

Most of the food we eat today was domesticated when people began learning to settle into agriculture. Today, we have tinkered plants, but only on the edge. Maybe we are getting higher output here, or there is resistance to plant herbicides. Nevertheless, in terms of converting wild plants into new elite crops, the progress is on. Macadamia, kiwi and vanilla are some examples of domesticated wild foods.

Today, there are less than 150 plants on which humans depend, and three grains such as rice, wheat, and corn account for roughly two-thirds of the globe's calories. Together with barley, they comprise around three -quarters of the world's cereal market. Overall, these crops aren't suitable for a rapidly evolving world. Modern humans have the opportunity to do even better. This is the attraction of domesticating an entirely new group of plants (Henry, 2010).

It is a tasking process to obtain crops from the wild or yield to consumers' demands. However, geneticists already have outcomes that may be showcased and tested. It's because modern day people have plant domestication tools which our ancestors could have never thought about: DNA sequencing technology permits scientists to accurately select the features they want.

Case Study: DeHaan Grain (Kernza)

DeHaan is a novel cereal variety which has been invented recently. It comes from a distant wheat relative, called intermediate wheatgrass. Contemporary wheat is generally an annual plant, and farmers must reseed the plant every year, however, DeeHan's grain is perennial. It has diverse growing seasons, meaning it needs less fertilizer (which consequently means less toxic runoff).

Our early ancestors chose annual plants because they usually produce more seeds and grow well in the disturbed soil and need to be replanted from the seeds every year so that they are easier to domesticate at the beginning. However, perennial plants do not waste energy and can reduce new rooting every year, which further means that farmers do not have to spend a lot of energy decomposing topsoil.

The famous prairie ecosystem that once cut across Kansas is dominated by perennial grasses. These plants are deep-rooted, tall, disease -resistant and grow on dense mats with no chance for weeds. They latch carbon in the earth and are ideal at handling unstable rainfall (due to climate change). But today, Namby-Pamby crop is planted on every acre of grassland found in North America. Other annual plants in high demand are corn, wheat and soybeans. This same situation is true for Brazil, Russia and China. Climate change makes this mission essential; while the genetic evolution is further making it practical.

As early as 1976, the Land Research Institute launched out to transform existing cereal crops into perennials. They first crossed regular annual wheat together with intermediate wheatgrass. The strategy worked, but only occasionally. Later, in 2001, DeHaan appeared. It was first grown on a tiny farm in Minnesota before spreading to other parts of the U.S. This was because of wheat hybridization and selective reproduction from wheatgrass (Henry, 2010).

Initially, inventor DeHaan believed that perennial wheat-like cereals required 50 to 100 years of manual strain selection to find the appropriate traits. However as genomic technology becomes quicker and cheaper, this work is accelerating. Breeders map the DNA of their seeds and apply genetic markers to choose plants with the desired traits. Genes that endow corn with low fragmentation may play the same role in wheat and potato beans. This technique is still conventional breeding, however not genetic modification, even so, DNA coding opens up shortcuts.

By 2010, what the DeHaan team had was brand new, very wheat-like-apart from its roots. The roots of wheat are very thin, extending only a few feet down. This new crop is based on a huge root system that extends 10 feet down, digs water deep underground apart from effectively drawing it from weeds. Deep roots also better absorb soil nutrients that may have been completely lost. Scientists have named it “Kenza” and expressed its usefulness in indigenous tribes and the field of etymology at large. The inventor says they don’t want any niche market that’s economically viable, but rather something that can considerably transform agriculture.

Since then, Kenza has attracted more and more collaborators, from universities such as (the University of Georgia, University of Kansas and University of Minnesota) up to the federal administration (USDA) to large companies (General Mills). Recently, both the World Bank and Gates Foundation have supported the project. This summer, Minnesota will harvest around 90 acres of Kernza land for Patagonia Provisions, an outdoors clothing firm’s new sustainable food production line. A brewery in Ventura,

California as well as another in Lawrence, Kansas have also experimented with this type of crop.

Despite this, Kernza still faces its own challenges. The seeds are rather small, the yield low and the shells stick onto the seeds, preventing the milling process. Researchers estimate that it will take roughly 20 years to make it perfect. Considering the domestication of wheat, rice and barley took 2,000 to 4,000 years, this actually isn't bad. Even after taking a few months inside the refrigerator, Kenza bread still makes the tastiest bread you can ever consume. It has a nutritious earthy flavor, indicative of rye and rye flour, but with the light texture of regular white bread (María de Cortes Sánchez-Mata, & Javier Tardío, 2016).

DeHaan flour can further be used to bake chocolate cookies, with half Kernza and the other part standard all-purpose flour. The DeHaan biscuits have a somewhat nutty taste, plus their complexity is quite similar to whole wheat, though there isn't any unpleasant chewing quality. It feels wholesome to eat Dehhan products since they are rich in fiber.

## 7.10 EMERGING TRENDS IN DOMESTICATION

These days, individuals are more probable to "prospect" to grab what a neighbor has domesticated (garlic, cashew, quinoa and mango) and grow it by themselves. This is what's called food slump because native plants are most likely to adapt to local conditions. The aforementioned potato beans make a good example. A Louisiana scientist called Bill Blackmon collected and bred the variety in the 1980s and evaluated the characteristics of well over 2,000 varieties. (This strain grows quickly; has rather small tubers, and tastes more wild.)

Especially for legumes, breeding involves gathering pollen from male flower parts with tweezers or a magnifying glass, before moving it onto the female flowers through the hand. Potato beans however don't tolerate this procedure well, since it resembles a complex, nearly flower-like orchid. All components involving pollination are concealed and uncooperative, besides the pollen only survives for a few hours per day. Therefore, when growing this wild food plant scientists rely on the goodwill of insects to pollinate chosen species in pairs, growing differently from other crops (Daniel Zohary, Maria Hopf, & Ehud Weiss, 2012).

Currently, the best potato strain produces roughly half of all potato plant tubers. However American honeybees have up to thrice the potato protein of similar weight, therefore they're more nutritious. Scientists also mention that the Apios Americana crop doesn't perform as great as sweet potatoes or potatoes as of now, harvesting also is becoming a problem. Potato beans develop on rather long subterranean stems (stolons), which require extensive excavation. Besides, being a vigorous vine, it not only supports the harvester well but other plants also.

The domestication of conventional old potatoes involves the selection of dwarf varieties with short stems. But recently, researchers have discovered genes that inhibit dwarfing in other plants like wine grapes and beans. With this information, scientists are attempting to secure down the appropriate gene in Apios, so as to separate it from the environment. It may take some years to perfect the plant gene, but work still continues.

At each harvest, researchers measure the proportion of aboveground matter (vines and leaves) to underground tubers, the amount of tubers plus how far separated they are from one another (more closely spaced tubers are simpler to harvest). They classify the plant DNA of every line and search for genetic markers that help to make the selection process easier. Among major crops such as corn and soybeans, thousands of strains with well-known and described features are stored and accessible to breeders. But for 'new' crops such as Apios, scientists must begin from scratch and find a workable solution (Daniel Zohary, Maria Hopf, & Ehud Weiss, 2012).

It is best to consider plant domestication as an adaptive procedure rather than a binary trait, which usually results in gradual changes without obvious stages. This procedure is very complicated. Scientists believe that successful domestication is reliant on many internal and external factors, such as life history, practicality, polyploidy, and a huge effective population. However, more cautious consideration of the comparative significance of these factors including how they work together can provide a practical basis for determining which crops may be ideal candidates for domestication, plus a better understanding of why domesticating certain crops seems to have hit a snag.

Although it may ultimately be hard to determine the origin of each functional allele, besides most of the work to date simply focused on alleles with greater impact, therefore it's believed that most characteristics are polygenic, and essential for domestication. Most of the mutations happen because the basic variant has detached the wild plant population. In case domestication is indeed done by adjusting to multiple genomes from standard genetic variation, it means that breeding instead of modification of genes might prove to be an effective way of domestication going into the future.

The experimental breeding of wild species domestication may be a difficult task and hasn't yet been completed. Because many plants have been domesticated independently many times shows that this idea is reasonable, although in most cases, the genetic flow between cultivation populations encourages multiple domestications. It might really be harder to domesticate new wild species from the beginning, partly because of the complex genetic pool of domestication traits, the restrictions of life history, as well as extension of the time required (Daniel Zohary, Maria Hopf, & Ehud Weiss, 2012).

However, some recently adopted crops, such as sugar beet, have shown great adaptability in the past few hundred years. This can be achieved quickly by carefully selecting candidate species and contemporary breeding methods (such as genomic selection), as well as high-throughput phenotype. Ultimately, better integration of the considerations mentioned above in plant domestication research shall help people to comprehend the adaptability of crops to the anthropogenic environment and help clarify the usefulness of research domestication as an illustration of experimental Evolution-Darwin advocated this idea around 150 yrs ago.

## 7.11 HOW TO DOMESTICATE WILD FOOD PLANTS

Nowadays, many commercial nurseries prefer to cultivate and hybridize. They domesticate and breed into garden plants with undesirable traits such as dwarfism, particular flower color, double flowering and uniform growth. These characteristics look good to humans, but they typically lack the genetic variety necessary for the reproduction or adapting to new environmental situations. In order for these characteristics to persist, nurseries need to reproduce these wild food plants asexually.

In the wild, the majority of plants reproduce sexually from seeds. Sexual reproduction leads to diversity between individual plants because any kind of plant has different adaptability in its ability to cope with ecological stresses (such as high temperature, drought, flood and other possible disturbances). Genetic variation can be a great strategy for plant species to adapt to future environmental circumstances, and seed reproduction can maintain the inherent genetic diversity of wild native plants.

*Cultivation of native seeds* Among the joys of growing native food crops is that there's no need for costly or complex facilities. Native seeds are ideal for sprouting in outdoor beds or flower pots. For most natives, germination outdoors is usually more improved compared to sowing seeds in the greenhouse, where extreme temperatures and humidity levels create ideal conditions for decay. Through outdoor propagation, when conditions for every species are ideal, the seeds shall germinate, a few in the cold early spring season, and some until summer when it's much warmer outdoors.

Seeds may be sown inside seed typical boxes or plastic pots measuring 4 to 10 inches in diameter and 3 inches or deeper and stuffed into the shade. For home gardeners, a small nursery with 10 to 20 pots may be built under a garden seat. Here it is undisturbed by dogs and children, shaded and looks neat. The exposed cold frame also provides a decent nursery area for seedlings and pots.

*Seed Sowing:* Native seeds can be densely sown (closely planted). Simply label with the appropriate name or date of sowing. A recommended rule of thumb would be sowing the seeds to depths similar to the seed's thickness, about 1/8 to 1/4 inches apart. The seeds are hardly covered. Following the sowing, conceal the seeds with some rough sand, which is ideal for potting soil because sand helps prevent the seeds from splashing in the rain. If the adjacent weeds are a cause for worry, consider covering the area with a decent spun poly covering like Reemay. Continue watering, typically every few days to 1 week.

*Waiting for germination:* Every native seed possesses its own germination schedule, which is rather different from the cultivation and selection of cultivated plants for rapid germination (such as annual flowers and vegetables). For wild plants, normally seed germination is variable. A few seeds germinate instantly within a few weeks, months, and even years, whereas others grow irregularly. This breeding strategy is beneficial for wild plants, because the offspring will be dispersed across time, which is a better approach for dealing with climate fluctuations.

Seeds that do not require pretreatment will sprout at any time within a week to some months after sowing. Different other species that require winter stratification will germinate after overcoming seed dormancy at a time suitable for each variety. Besides, flats that are not able to germinate originally will typically germinate the subsequent year. Don't discard them out, but instead be patient. The wild plant seeds can be propagated in autumn or winter, then placed outdoors without any artificial indoor cold stratification (refrigerator) needed for planting certain species in spring, it's the easiest germination method so far discovered.

*Seedling transplanting into larger landscape pots:* In the initial growing season, many local seedlings can be left in the original unit. In case the seedlings become overcrowded, they can be mildly separated and potted during the first summer. Contrarily, you can wait until the subsequent spring to relocate the plant seedlings to a fresh location. Don't forget labeling the plants. During transplantation, 3 to 10 seedlings are mixed per pot together (except for tree species). It ensures more than one crop; the new planting shall have a certain genetic diversity and shall be able to reproduce viable seeds. The liquid seaweed fertilizer diluted once every two weeks will keep the seedlings in good condition and strong.

*Overwintering seedlings and nursery protection from rodents:* Sprouted pots and germination pans require winter protection from harsh weather and strong winds, just like providing continuous snow cover to the garden. Multi-layer winter- grade Reemay crop concealed with white plastic functions quite well. The plants must be frozen prior to covering, otherwise, rodents might choose the nursery that you cover since there are ideal nests there. Make sure to place the mouse trap below the cover to provide more protection.

*Preparing a seed incubation bed:* The key advantage of a germination bed is the fact that it doesn't need much watering, no more than a typical bungalow or pot plant, plus the seedlings can germinate even before transplantation. For woodland species, choose a fully shaded or partially shaded location, but for most other types, consider a shaded location, but for plants that require dry, sunny conditions, choose sufficient sunlight.

Preferably, cut the frame into a rough shape (without pressure treatment, it will release any existing harmful chemicals). A 4-foot-wide by 10- foot long flower bed can hold many seedlings. You may further use logs measuring from 6 to 10 inches in width. Add some thick sieves at the bottom, mix them over and then fit them into place. Add some weed-free compost mixture. You can develop your own growth medium through mixing at least 3 portions of weed-free manure or leaf mold, a single part of vermiculite and single part of rough construction sand.

Compost and leaf mold shall contain many beneficial microbes and will slowly release nutrients across time, which is different from the case where sphagnum-oriented potting soil is antiseptic and lacks nutrients. The seeds may be sown horizontally in 4-foot rows on the bed and labeled using durable plastic stickers. If there are sources of weeds nearby, like old fields or dandelion-filled lawns, consider covering the plant bed with some Reemay.

*Germinating Edible Ferns from Spores:* Ferns are rather ancient plants whose ancestors originally appeared on earth around 300 million years back. They form a group of primeval plants, which ruled the entire earth before flowering plants emerged. In the age of dinosaurs, ferns and different other primitive plants (such as club moss and horsetails) reached amazing proportions, many of them exceeding 100 feet in height. This duration of the earth's heritage experienced a global climate with warm temperatures and elevated humidity conditions, which is an ideal condition for ferns to flourish. Today, edible wild ferns can actually be domesticated, such as the ostrich fern fiddlehead.

Ferns possess a unique propagation strategy that involves two unique stages in the life cycle: prothallia and mature ferns. At some point during the growing period, mature ferns release spores, the plant's sexual reproduction cells. With enough moisture and light, the spores will start to grow and develop small flat plants, commonly known as prothallia, which is the second stage of the life cycle. Both male and female reproduction organs mature on prothallia. In case fertilization happens, the egg cells will grow into sporophytes (young ferns), while the life cycle of the new ferns will start again.

Some fern spores can multiply indoors in bright windowsill under direct sunlight or from light. You'll require some antiseptic, peat-based potting earth that has been wetted and cooled with boiling water, water in an atomizing bottle (which has also been boiled or cooled), including a new transparent plastic deli box. The spores are sown in the container, plus if not sterilized, algae or mold may overrun the place. Naturally, ferns will germinate in moss, rotting wood, or humid bare soil (such as near the stream).

Fill the transparent plastic container half full with water, and then quickly swap the lid. After preparing the fern spores, open up the lid and sensibly sprinkle a small amount of spore powder on the soil surface, the same as when physically seasoning with some salt. Water gently, then place the container inside a warm, bright place, away from direct sunlight or below the lights. Check germination for the next few weeks or longer. But don't open the container not unless you believe the soil has dried. In this case, use a sprayer to lightly water.

Initially, the fern germination resembles a green coating found on the surface of the soil. The next stage is the growth of a small green tablet-like structure called prothallia; it's when sexual reproduction happens. After around a month or so, you'll see some fern leaves. It means that you've effectively transferred from the reproductive stage to the young fern. Here, you must take out the cover. Water your seedlings regularly, but don't over-water them, otherwise, the plants may rot. Eventually, it will be necessary to use rather mild algae fertilizer to provide fertilization. Meanwhile, small ferns measuring around 1 inch tall, maybe repotted within a decent organic compost-style potting soil. During early summer, transfer the growing pots outdoors into some shade to regulate their wild growth pattern. These can then be planted back in the garden during late summer and early fall.

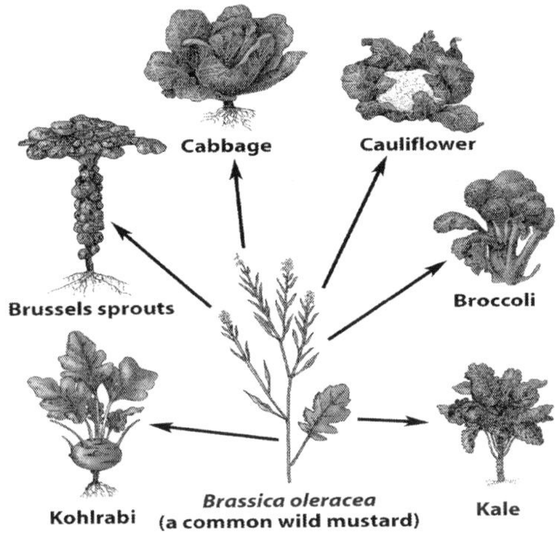


**Figure 7.3:** The wild mustard was used to produce modern-day plants like kale and cabbage.

*Source: https://www.thinglink.com/scene/761585201867390976*

The management of the man-made environment and species is closely related to the ecosystem and evolution of food plants collected by people. This is indeed the case with wild food crops, which exist in the continuous management spectrum of plants. In view of the alarming rate of deforestation affecting human-based ecosystems, wild food plants are increasingly collected, and farmers keenly manage these species so as to ensure their viability and accessibility. Research shows that 67% of "wild" food plant species reported are well managed by humans, and nearly all plants that have been managed are finally transplanted. The most common flow of transplanted wild crop comes from forest areas to agricultural fields where they are domesticated. This usually happens in rural areas that border the forests.

On the contrary, the most significant reason for not transferring "wild" plants is the long growth time. It's also worth noting that well over half of all "wild" edible crop species are classified as weeds within the scientific community. Ultimately, based on the management procedure and apparent richness, "wild" edible crop species can be subdivided into multiple exclusive groups. Management of 'wild' food plants, including the administration of species mentioned as weeds in some scientific literature, constitutes a vital adaptation approach for farmers to ensure their general food security in case of heightened deforestation.

Rural households collect wild edible plants from vastly intervening environments (such as agricultural fields), higher living standards (such as family gardens) and less intervening areas (such as forests). However, due to the shocking loss of wild habitat, they are increasingly collecting wild food crops from man-made ecosystems. For example, it is documented that families farther away from the forest (that is, due to extreme deforestation rates) prefer gathering closer to their residences, which led to coining of the word "botanical diet paradox", which shows that farmers are becoming more dependent on agricultural "wild food plants" when forest areas decline.

# 8

# *Ultimate Foraging Guide for Edible Wild Plants*

## CONTENTS

## 8.1 INTRODUCTION

With the ever -increasing worry and need for food security to ensure that the world's population is maintained, scientists and governments have looked into the possibility of making wild plants as a source of food. In Africa, native cultures made use of the resources readily available to them as a source of food. They made use of a number of wild plants as food. A method they acquired food is through the gathering of wild plants such as wild mushrooms. Till today, there are people who feed on wild plants. For most of the developing countries with most of the populations living under the poverty line, the use of wild plants as food is a cheaper method of acquiring food (Ballerini, 2012).

In the modern world, most of the food is gotten through the planting of the food stuffs. The option of utilizing wild plants is quite sensitive. There is the presence of a number of challenges one of which includes the possibility of most of the wild plants being poisonous and identifying those that are good for human consumption. There is still a high regard for wild pants as they present the advantage of human beings able to get nutrients from them. In modern society, some people still value the wild plants for the nutritious value they possess and the fact that they are a source of food. There are a variety of wild plants that grow on their own.

It is crucial however that individuals know a system that can help them identify which plant is edible and which one is not. This is in order or them to avoid cases such as stomach aches or poisoning. Most individuals have basic knowledge of some of the plants that are edible (Ballerini, 2012).

## 8.2 HISTORY ON WILD PLANTS FORAGE

It is known among the different communities that wild plants have been used as food. This was long before there was the planting of plants for food. Initially, our ancestors hunted and gathered food. They made use of the resources available in the surrounding for food. Over the years, there is still the utilization of these plants for their nutritious value among other reasons. There are various parts of the wild plants that are utilized, which include their roots, leaves and fruits amongst others. Some of them include mushrooms, algae, fleshy fruits and berries among others. Some individuals have gathered and eaten these plants based on the knowledge they acquired from their ancestors.

For one to properly collect the right plants, they need to have knowledge of things they should look at so that they do not end up picking poisonous plants. Some of the wild plants are mostly categorized as weeds. Weeds are plants that naturally grow in the fields and most people may refer to them as unwanted plants. In recent days, there has been the acknowledgment that these plants are more nutritious compared to other domesticated plants. Scientists have revealed that some plant species have genes that can be useful during research. There are a number of wild plants that are known among the different communities (Ballerini, 2012).

## 8.3 BENEFITS OF FORAGING EDIBLE WILD PLANTS

There are a number of advantages of the wild plants. The advantages may be in comparison to the use of domesticated plants. Some of the advantages are as follows:

### 8.3.1 They Are a Source of Free Food

There is no doubt that the use of edible wild plants is a cheaper way of acquiring food. All the individual has to do is to get to a given field. Collect the given edible plants and prepare a meal for their family and friends. This method is cheap with the fact that the plant species are freely given to people by mother nature. Most of the plants reoccur in the environment. Some of the wild plants are weeds. A good example is pigweed. This type of weed may grow near homesteads making it accessible to the population. Getting these plants is cost- free, it may only require that the individual sets aside time to go look for the plant. Time may be the expense in this case.

Some of these wild plants are usually collected by farmers and some of them sell these plants in the supermarkets and grocery stores to make them more available to individuals who live in the city. Though they are sold in the supermarkets, they are usually cheaper compared to other plants. This makes it more affordable to most of the world's population. This factor has been beneficial more so for the developing countries which have most of its citizens living under the poverty line. The eating of wild plant has been one of the ways in which people living under this line can acquire food and fend for their families.

Apart from these plants freely provided, it is a cheap method of getting nutritious foodstuffs. Currently, highly nutritious plants are highly expensive because of the value they pose to improve the health of the individuals. This is a cheap way in which they can get nutritious food. This means that they can get healthy food but a cheaper price or no price at all if they can get these plants on their own.

### 8.3.2 They Are a Source of Fresh Food

These plants naturally grow in the environment. They make use of the resources provided by nature to grow and flourish. This is a healthier approach to eating fresh food. First of all, the individuals can harvest the plants individually in the fields making these plants fresh as they have been recently acquired. Secondly, they are a fresher approach to domesticated food types. For most of these plants, they have chemical components that have been accumulated as the plant grows. The fertilizers that are used are not a healthier approach to humans as these chemicals may accumulate in the body. Some individuals have also laid out concerns about the chemicals that are used in these plants.

This has caused most of them to lean towards eating of wild plants. They are a more wild approach to food. They do not have any chemicals as they grow wildly.

### 8.3.3 They Are a Source of New Flavors

These kinds of plants are usually unique more like the unique tree species found in the forests. They provide a new flavor. Most of these plant species are not usually found in stores or in forests. Most of the well-known restaurants make use of these plant species so that they can achieve new flavors. They make recipes that are able to neutralize the toxic effects of these plants making it the same for people to consume. Most restaurants earn large with these new flavors. Some of these wild plants can also be utilized as spices in food. They give the food a new taste that is unique. The wild foods also present a new taste in fruits and a said to be quite delicate. A good example may be the wild pawpaw. These wild fruits are said to be so delicate that transporting is quite delicate. The use of these wild plants gives a new perspective to food taste that is unique. For restaurant owners, the use of these plants allows them to make new dishes that are unique to the restaurant it was created in.

### 8.3.4 It's a Chance to Get Outdoor Exercises

These wild plants are located in the fields. For one to access them, they have to go to the wild outdoors to look for them. This is the option usually taken by individuals who prefer to get these plants on their own instead of purchasing them in the supermarket. The picking of these plants may involve hiking activities. They may climb mountains or go to the valley as they look for these wild plants. Most of them choose to hike in areas where they have had knowledge that there is the presence of the certain wild plants they want. By walking to these places, their bodies get to exercise. There are other activities that are conducted which contribute to them exercising. Some of the activities include when they stretch to pick a fruit for instance, a berry. The stretching allows the limbs to be more flexible. There is also the bending as they try to collect the root or leaves of certain plants.

There are some individuals who choose to pick these wild plants early in the morning or late in the evening. When they conduct these activities at these times, the sun shining outside is usually a good source of vitamin D and it is also a form of a wild workout. The conductance of these activities allows individuals to get a fit body without them having to go to the gym. It is also a good way for them to detox from the busy life schedules and they are able to relax.

### 8.3.5 Great Source of Nutrition

For most of the plants that have been cultivated, they have been sprayed by chemicals. The chemicals have led to the modification of the nutritive value of the plants in that the level of nutrients has reduced. It has also modified the taste and nutritive value. When placed in comparison to the wild plants, the wild plants are highly nutritive. They are also highly nutritive compared to the foodstuffs sold in the supermarket. Researchers have been able to establish that these plants are highly nutritive and they conducted a comparison. They compared wild greens dandelion to the ones that are grown and sold in the supermarket. The result was that the wild ones were found to be highly nutritive: seven times more than the ones sold in grocery stores and supermarkets.

Even for the wild fruits, they are found to be more nutritive than the ones that are planted in farms and gardens. This was proven by the research that was conducted on the phytonutrients found in apples. The research revealed that the phytonutrients found in the wild apples were highly nutritious to a greater extent when placed in comparison to the normally planted and sold apples. On a side note, an individual can also gain nutritious value as they pick the apples in the fields. This is because the sun is shining on them and they are able to get vitamin D as a bonus.

## 8.3.6 Helping Ensure Food Security

All around the world, there has been a global problem of lack of food security. This implies that the country's food supply is not able to meet the demand from the population. This may be due to the fact that the population is larger than the amount of food that is planted. The option of using wild plants as a source of food has been one way that the government is able to maintain food security. Wild plants being used as food has been able to ease the pressure that has been placed on the need for planting enough plants. The advantage of this type of food supply is that it is not dependent on artificial fertilizers and irrigation to ensure that the plants grow. They usually make use of the wild resources provided by nature in that even though the seasons have changed, these plants are able to plant. The added advantage is that some of these wild plants need a short duration of time for them to grow. With the short growth life span, these plants can be frequently harvested and it will take a short time for them to regenerate ensuring that there is a continuous supply of these plants. The other advantage is that there are a number of wild plant species that can be utilized as food. This provides people the option of being able to pick the type of plant they would want to eat.

## 8.3.7 Enables People to Connect with Nature

It is really important for people to connect with nature so that they can understand nature more. This will allow them to conserve the environment with more understanding and more sensitivity to the environment. The harvesting of wild plants is one way in which people can connect with nature. As the individuals are harvesting the plants, they are able to connect to nature. They get to know the cycles in nature that people may not have experienced due to the ever- changing climate due to global warming. There are different plant species that grow in different seasons. There are wild plants that grow during summertime. Some of the wild plants include the wild raspberries, juneberries and persimmons.

Some of the plants that grow during spring include the watercress, morel mushrooms, dandelion greens and ramps. This is a good way in which individuals can notice the ever -changing seasons through the plants they are harvesting. Also, individuals get to connect with nature as they pick the wild plants with their own hands. It allows them to experience nature from a different point of view.

### 8.3.8 Sustainability

When looking at sustainability, it involves one looking at the point that what they expect is what they will get. With the case of wild plants, there is the sustainability aspect that is demonstrated with the fact that one knows they are eating healthy and fresh food. They are naturally organic and fresh. This is indeed true as they do not contain any chemicals in them as they are not sprayed by any chemicals or pesticides. These plants make use of resources provided by nature such as the rainwater and the nutrients provided by the soil. They are also sustainable in that they can be easily harvested. This implies that it does not require any heavy machinery to harvest these plants. This is practical as individuals can pick the required plant. If it is wild berries, they can easily pluck them like the normal berries.

### 8.3.9 It is fun

Fun comes in the aspect of foraging for these plants. The activities that are involved in the process of foraging of food. Take for instance hiking. This involves individuals going out in the open field to acquire these plants. If these activities are conducted along with family and friends, it makes the experience more fun and exciting and there are memories that are made. It is also a good way for people to pass the time and relax while being surrounded by nature and the wild atmosphere. If children are involved in the process of picking these plants, it is a way of sparking interest in them. It will engage them and it will be a fun experience as they get to bod with both their parents and friends. It is also a good way for them to acquire knowledge on these plants and they can get a guide on how they can identify the edible wild plants.

## 8.4 HAZARDS OF FORAGING EDIBLE WILD PLANTS

This includes some of the downfalls of looking for edible wild plants. In as much as there are many advantages of foraging for these plants, it is also important to note its downfalls more so for individuals who do not have any experience when dealing with this sector. Some of the hazards are as follows:

*Potentially damaging of the surrounding:* In as much as harvesting of these plants is advantageous as it provides food, there is the danger of damaging the environment through overexploiting these wild forests. The same case was experienced with forests in that the demand for wood led to the overexploitation of forests. The same case can happen to these forests in that people can over harvest these plants. The challenge is that there may be no indication of how much plants individuals can harvest without exploiting the environment. There are cases where people are overenthusiastic about foraging these plants and end up harvesting all the wild plant species. This has made room for other plant species, which may not be edible to grow in place of the initially existing ones (Saunders, 2012).

There is also the possibility of foragers destroying the environment in a number of ways. If the forager is inexperienced, they may harvest the required plant part in the wrong manner which may lead to the significant damaging of the plant. Also, most of these wild plants grow in areas that have undergrowth. As the foragers are harvesting, they end up stepping or trampling on the other vegetation leading to the significant damage of these plants. There is also the possibility of them crushing other plants as they are harvesting the edible plant parts. As they are picking the plants, they may end up destroying the wild homes for animals that reside in these plants or in the surrounding. For people who do not have experience in foraging, they may not have the ability to identify the areas that are highly vulnerable to destruction.

There is a significant possibility of the destruction of the environment more so leading to the depletion of these plant species in the case where these plants are fully taken up as an option in ensuring food security. The pressure of meeting the population may lead to the over- harvesting of these wild plants.

*Possibility of being poisoned:* Though there a number of plants that are said to be fit for people to eat, it is important for one to note that there are some of these plant species that are poisonous. When foraging, it is important for the forager to have knowledge of the plant species and being able to properly distinguish between them. If the forager does not have knowledge of these plants, then there is a high possibility of them picking poisonous plants. This is because, in the case of the wild plants, there are poisonous plant species that may look similar to those that are safe for consumption. This mistake can also be incurred by experienced foragers. There are usually some features that distinguish between the poisonous and edible plant species. These features can be missed even by experienced foragers.

There are known cases of some forgers who succumbed to the poisonous plant species. A good example is Christopher McCandless who is well known for his writings on edible wild plants. The writer died after he consumed a wild plant that was poisonous and it killed him by weakening his immune system. The challenge, therefore, is the forager being certain that the plant species is not poisonous at all. The wild plant can also be harmful if they are not properly cooked. Usually, there are special recipes that are utilized fitted for the given plant species. There are some of these plant species when undercooked may end up being highly toxic. This may require that they are cooked for a longer time to reduce the toxicity. There some that need to be cooked with other elements to neutralize the toxicity of the plant. There are those that are to be boiled before they are cooked with other foodstuffs (Saunders, 2012).

*Unfamiliarity:* For the foragers that do not have much knowledge on these plants may end up incurring damages such as stomach poisoning. When dealing with unfamiliar grounds, they may not be able to identify which plant species is safe for consumption and which one is not. It is generally known that most of the wild plants that are edible are bitter or tough if the technique used to prepare the plant is not proper. The unfamiliarity may come about the fact that one may not know the degree of boiling that is required. One may also not know the extent to which the food should be tender. In summary, the lack of recipes that should be followed during the preparation of these wild plants. This may lead to the possibility of the consumption leading to a stomach ache or even the poisoning of the individuals consuming the food. The unfamiliarity may also be indicated by the fact that some of the foragers may not know for how long most of these plants can be kept without them going bad. There have been cases where the forager harvest the plant species but they do not know the method in which they can utilize these plants so as to make a meal out of them. This may lead to the wastage of the harvested plant species.

*Arrests:* Most of the foraging activities are usually conducted in open fields. However, there is the possibility that one may be foraging but in a piece of land belonging to a certain individual. The law states that foraging in another person's property is illegal. If an individual forage in another person's lands, they may be trespassing and end up being arrested. In some states, there are laws that have been set with regard to foraging. There are states that forbid foraging in parks. There are those that allow foraging in the case where the individual has no other way of fending for themselves. Some of the laws limit the amount of plants that one can harvest. Though this is so there is the challenge that the limitations on foraging may not be quite clear. This has contributed to the cases that have been incurred by people foraging in areas where these activities are not allowed. This may lead to a number of foragers being arrested. It is also a challenge for the foragers who have to fend for large families in a state where the number of plants that can be harvested has been limited in that if they harvest beyond the limit they will be arrested (Saunders, 2012).

*Contamination:* People who carry out foraging prefer these kinds of plants as they are more wild compared to other plant species. Even though this is so, there is the possibility if these plants being contaminated. Usually, before an individual conducts foraging activities, they look into a number of factors such as the possibility of the location of these wild plants being sprayed with chemicals. This is if the field is located next to a farm. They also look into the possibility of the location being next to any pollutants. For instance, if a river flowing in the given field is polluted. If this is so, there is the possibility that the individuals are consuming poisonous chemicals absorbed by the wild plants.

For the forager, there is the challenge of certifying that the land has not been sprayed by chemicals or pesticides. It is also a challenge in ascertaining that the land is not polluted by any pollutants. This poses a significant risk to the health of the forager more so if the land is polluted by radioactive material. It may lead to poisoning or death to the furthest extent. It is therefore important to be keen on identifying the proper location for foraging.

*Possible attacks:* Most of the foraging activities are conducted in fields. Some of them are conducted in forests. The activity is done depending on the location of these plants. Some are located in dangerous locations. For instance, if they are conducted in forests there is the possible threat of the people being attacked by wild animals. There is the possibility of being bitten by snakes or by other wild animals. This is a risk to the forager who may end up being injured or even worse they may end up being killed, which is the worst- case scenario. If the forager does not have information on the locations that are safe and which one should be kept out of bounce they may be putting their life in danger.

Overall the dangers posed by to foraging of edible wild plants generally revolve around the uncertainty that it encompasses. The uncertainty is the problem that may lead to the poisoning of individuals and potentially killing them (Auerbach, 2011).

## 8.5 HOW TO IDENTIFY EDIBLE PLANTS IN THE WILD

There are a number of factors that come in to play when one wants to group a plant as edible or non edible. Some of these factors include whether the plants are poisonous or not. If the plant is poisonous, then it is not fit for consumption while those that are not poisonous are the best plants for consumption. Usually, the toxicity of the plant is a major factor but there are other factors that can come into place such as the amount of cellulose in the plants. There is the level of cellulose that the human body can handle. This is with regards to the digestibility. The plants usually harvested by foragers are usually low on cellulose - the lower the cellulose the higher the rate of digestibility. This implies that large amounts of cellulose may be indigestible by human bodies (Auerbach, 2011).

It is important for the forager to have a guideline to help them identify which plants that are edible and the plants they should avoid. The identification process is a crucial process so that one does not end up picking poisonous plants. For new foragers, there has been a method that has been utilized over the ages and it is referred to as the universal edibility test. This test is mostly used by people who may not be sure about the plants they are picking whether they contain any poisonous elements. This test involves the individual getting in contact with the chosen plant species.

The process involves them ingesting small bits of the plant. This is done over time so that they can gauge whether the plant has diverse effects on the person ingesting the plant type. The procedure is usually conducted on the part plant to be ingested. The following steps are undertaken to ascertain the plant is fit for consumption. First, the forager is required to smell the plant part. This is defined as taking a strong sniff of the plant part. If the plant has an awkward smell, then that is a red flag. If it also smells bad or there is a rotting smell that is an indication that the plant is not fit for human consumption. The smell of the plant can be utilized as a great indicator of the edibility of the plant part (Department of the Army, 2009).

The other step may include the forager squeezing the plant part on their elbow. The juice produced by the plant can also indicate whether it is fit for human consumption. If the juice causes an irritation on the skin, then it should not be consumed as it is poisonous. If the plant does not cause any irritation on the skin, then the forager can proceed to place the plant on the lips for about 15 minutes and if there is no irritation on the lips then they may proceed to bite the plant. It is advised that the bite taken should be relatively small. If the plant tastes soapy or bitter, it is advised that the individual spits out the plant. If the bitterness is bearable then the individual is advised to hold the plant for a given amount of time let's say 15 minutes. If the plant as passed through all the steps and has proven not to be poisonous then the plant is fit for consumption.

Under the edibility test, if the individual is unable to properly identify the plant, then there are factors that should be taken into consideration. The new forager should stay away from plants that tends to be milky or produce a discolored sap when they are cut. It is also important for them to keep away from plants that have a soapy or bitter taste when the forages sample the desired plant part. They should also avoid plants that have thorns, spines or fine hairs as most of them tend to be poisonous. They should not consume plants that have a growth pattern of three leaves and they should avoid consuming seed or bulbs from the various plants. The above characteristics are commonly found amongst plants that have toxic characteristics.

In the identification of plant parts, it is important for the individual to have some basic questions. There is a need for the individual to have an idea of what plant type they want to pick. They should also know the place where they can find this plant species. They also need to know the mode of preparation for these plants so that they do not harvest the plants only for them to go to waste. In the case of knowing the type of plant, they want to pick there is general knowledge of the common plants used in foraging. There are plant parts that are commonly utilized in the case of wild plants and they include the roots, leaves or stems. There a number of plant foraging that are utilized and well known. They include fruits, nuts, roots, greens and mushrooms (Department of the Army, 2009).

It is also important for foragers to know that not most of the pictorials on plants that can be consumed are fully utilizable. With the use of pictorials, there is a large possibility that an individual can miss out on some distinguishing factors hence making them pick plants that are poisonous. This is because there are plants that are known to bear a resemblance to other plants in various seasons. The non-edible plants may bear a resemblance to the edible plants. It may be important to incorporate other factors such as the smell of the plants. Understanding the seasons may be a crucial piece of information that is used in the identification of plants. As said earlier, the wild plants are known to grow in certain seasons. There are those that said to grow in spring while others grow in summer.

A good time for identifying the edible plants is through the knowledge of the type of flowers they bear. There are plants that bear five- lobed flowers while others bear three lobbed flowers. The color specification of the plant may also matter. The identification through the use of flowers is a method mostly used by experienced foragers.

## 8.6 FORAGING ETHICS

The main aim of foraging ethics is to ensure that even though these activities are being conducted, the environment is safeguarded. It is also there to ensure that foragers do not harm themselves by picking the wrong plants. Foragers are usually encouraged to forage ethically this is done to ensure that the same plants are present for the future generation through the responsible harvesting and conserving of the environment. Under foraging ethics, it can be narrowed to respect. The respect may come different ways that are demonstrated as follows:

### 8.6.1 Respect for the Plant

Usually while foraging, the foragers are encouraged to harvest in a manner that ensures the plant is sustained. This will allow the plant to grow in a number of areas. To respect the plant, one needs to have knowledge of how they can harvest the required plant parts so that they unsure the plant is not damaged. There are other ways one can respect a plant, which includes the forager conducting a study on the area in question so that they are able to identify the plant species that are endangered by identifying these species of plant, they will be able to avoid the plants preventing them from going to extinction. They will also be able to avoid a lawsuit as in most of the countries harvesting of endangered plants is illegal.

Foraging ethics requires the forager to be conscious of the plant when harvesting in that even though the plant bears a lot of fruits, the forager will only harvest the amount that is required. It also involves them utilizing a proper method of harvesting so that the plant does not end up being damaged. For example, it may be more ethical to use a sharp knife to harvest the leave rather than tearing the leaf from the plant. Foraging ethics requires the forager to ensure that the tools used in the process of foraging do not lead to the plant getting diseases. This may involve the forager conducting activities such as sterilizing of the tools. Also in the harvesting of the plant with regards to the roots. By harvesting roots, it is important to know that by removing the roots, the plant gets damaged. Therefore, the forager should ensure that they harvest the roots if need be. The other condition is if there is a surplus amount of the given plant. It may be ethical for the forager to replant the seeds of the given plant.

### 8.6.2 Respect for Future Visitors

It is among the foraging ethics; it requires that the forager harvests what is enough for them even though the land they are foraging in belongs to them. They need to be conscious so that some plants remain for future visitors. It is definite that other foragers may visit the same field later on. Apart from harvesting in a conservative manner, they should also leave the environment in a good condition that will be presentable for future visitors. The forager should pick up their garbage and properly dispose of the litter.

### 8.6.3 Respect for the Law

For most of the states, they have laws concerning foraging. In some states, an individual cannot forage if they do not have the required license. There are laws that also limit the amount of plants that can be harvested by the forager. There are laws that indicate the places that can be utilized for these activities. These laws also indicate the penalties that will be imposed on individuals who break them which may include them serving jail terms or community service. It is therefore important for the forager to be aware of what is required by the law.

### 8.6.4 Respect for the Ecosystem

The forager needs to have an understanding of the ecosystem in that they are aware that the given plant plays a given role in the ecosystem. There is usually a symbiotic relationship between plants and animals and the forager needs to know that if they remove certain plants they will damage that ecosystem. By getting an understanding of this, they will be able to harvest with much understanding.

# 9

# *Factors Affecting Wild Edible Plants*

## CONTENTS

## 9.1 INTRODUCTION

The use of wild edible plants has grown over the years and is significant in the 21st century. There are a number of factors that undermine the use of these plants. These factors play a role in the significant use of the edible wild plant amongst people in the current day. The factors may vary according to the state the edible plant is used. Some of the factors may be related to the laws that have been placed with regard to the use of edible plants. The laws also play a role as a factor affecting edible wild plants. Among the factors include social, economic, institutional, cultural and environmental factors. All these factors have a strong interrelationship with wild edible plants by influencing the level in which these plants are used amongst people. We also try to look into how these factors influence the management of these plants with the main aim being conservation of these plants so that the future generation is able to make use of these wild resources.

## 9.2 INSTITUTIONAL FACTORS

The institutional factors are also known as political factors. The government is significantly involved in how these factors will influence the wild edible plants. These factors are related to the government and the role it plays in governing the use of these edible plants. In most of the countries, the governments have been faced with the challenge of trying to ensure food security. With the increasing demand in the population, they have faced quite a challenge. They have opted to look for other means to sustain the ever ranging population. They have tried to increase food production by putting more land into agriculture and trying to support the farmers in different ways. Even though these efforts have been made, there is still a significant problem of food security. The fact that there is a preference for food that people would like to consume makes the problem difficult.

Among the methods the government is curbing the issue of food insecurity, most of the government has initiated efforts to put more land into agriculture. By so doing they will affect the wild habitats of the wild edible plants by destroying them. This will cause a reduction in the numbers of the wild edible plants by destroying the plants. This leads to these species of plants not being available for future generation.

## 9.3 LAWS ON HARVESTING OF WILD EDIBLE PLANTS

With these issues, the government has looked into the use of edible plants among its citizens. In most of the countries, the government allows the citizens to make use of the wild edible as a means of sustaining the ecosystem. These plants are usually readily available to the citizens as they can readily access the plants. Some of the governments have allowed these plants to be sold to the citizens. It has proven to be effective. This has been demonstrated by most of the countries in sub -Saharan Africa where most of the populations make use of the wild edible plants. The plants have been used in facilitating food diversity for the citizens. It is also a means to ensure that the population can access highly nutritious food as the edible plants have been said to be highly rich in nutrients and micronutrients. The added advantage being the fact that it improves the taste of food. It has mostly been used by poor families who have no means to support their families. They are able to acquire the plants to provide for their families while some of them purchase these plants as a means of generating income. This affects the wild plants in that most of the plant species are being consumed and the likelihood may be that most of these plant species may head to extinction.

Institutional factors affect the wild edible plants in that they will influence the availability of these plants for the future generation. For some of the countries more so the developed countries, the government is knowledgeable on the need for the conservation of the plant biodiversity in that they have set laws that influence how these wild plants are harvested. For most of these countries, there is a limit on how much of the plants can be harvested by foragers. There are boards that have been put in place to control the harvesting of these plants. The decision on the amount is made after research has been conducted by the boards in charge of the wild edible plants. They make a conscious decision on the appropriate amount of plants that can be harvested by the individuals so that the species of the plant can be maintained for the future generation.

Besides the limited harvesting of plants, there are laws that have been put in place that may require the replanting of these plant species. For instance, with the harvesting of the wild edible plants for their roots, it leads to the significant damaging of the plant species and if this is not controlled then it may lead to the significant eradication of these plant species. Some governments have made it law that for foragers harvesting the roots, they are to replant some of the seeds from the plant or roots from the plant. By so doing, they will be able to ensure that these plant species will not go to extinction as the plants will be rejuvenated.

For some states, they have prohibited the harvesting of plants through the means of uprooting. This means that the harvesting of plants for their roots is prohibited in some of the countries. Some of the plants under this category may include lichen, fungi or algae. For most of the countries that have put these plants, they have enforced these laws in that if an individual is caught breaking the laws are likely to face imprisonment or paying a hefty fined. This affects the wild plant species in that they are utilized but there is proper management when using them ensuring that they are carefully used hence the different plant species are conserved.

There are also some laws that have been placed restricting the harvesting of certain species of wild edible plants. This is usually done in the case where these plant species are limited in number or are considered to be endangered. There is usually a list indicating the plant species that can and cannot be harvested. For these wild edible plants, they are covered by acts that deal with endangered species of plants. These laws also limit the owner of the land in that he or she cannot harvest these plant species. This affects the wild edible plants in that they will be sustained and can be used by the future generation. There are some countries that have prohibited the harvesting of these plants for commercial uses. For these countries, the only condition that allows an individual to harvest these plants is if they do not have any means to provide for themselves or their families.

In as much as there are countries that there are laws controlling the use and management of wild edible plants, there are those that do not have laws. For instance, in most of the developing countries, there are no laws on foraging. This negatively affects the wild edible plants in that there will be extreme harvesting of these plants. Overenthusiastic foragers may harvest large amounts of the plants for the joy of it or to experiment on them. In the countries where large numbers of the population live under the poverty line and there are no laws controlling the harvesting of these plants, there will be extreme harvesting of these plants so for both domestic and commercial purposes and the impact will be felt later on when these plant species will no longer be there. A good case scenario that can demonstrate the implications of the over-harvesting is with wild forests. With the wild forests, there has been harvesting so as to provide wood to make furniture. This has been done to extreme effects that it has led to desertification for some of these areas and has negatively affected these areas and the same may be experienced with forests.

The institutional factors placed by the government through laws significantly affect wild edible plants with regard to their use. The presence of these laws positively affects the use of wild edible plants in that there is proper management with regards to their use in that there is controlled use of the plants. There is a certain amount of the plants that are harvested and also the plants get time to regrow and replenish their amounts in the environment. It also ensures the protection of the endangered species of plants as there are laws that champion these endangered plant species. With these laws, they are fully effective if the government is committed to ensuring that the lawbreakers are punished. The citizens also play a role by working beside the government ensuring that they have a common goal in ensuring that these plant species can survive for many generations. The laws are usually improved or modified according to the situation at hand. The general impact being there is a positive effect on the edible wild plants with regards to its use and management.

## 9.4 LAWS ON LAND

Plants grow on soil hence land. The area or piece of land where the edible wild plants grow plays a crucial role with regards to whether or not it will be harvested. The condition of the land will also play a role in knowing whether it will be harvested or not. The reason being that most of the human activities involve the use of land. The activities will determine the nature of the land. The use of wild edible plants has been affected by the laws placed by the government. There are laws that have restricted the foraging for the wild edible plants. There are laws that have been enforced to protect the fields bearing these wild edible plants. Most of the wild plants grow in forests, in agricultural fields or bare lands.

Some of the laws include the restricting of foragers form conducting their foraging activities in private lands. Private lands are those that belong to a certain individual. The law requires that even though some of the edible plants are located in the given private land, the forager is not to forage in that land failure in which will lead to the individual being charged with trespassing. If they are to conduct these activities, then they need to seek consent from the owner of the land. If the owner gives the consent, then they can conduct the activities. If the owner refused then they are to obey the decision made by the owner and look for other lands to forage.

The law was set to protect the privacy of the landowners, to ensure the safety of the owner as well as his or her property and ensure that there is respect for the environment. This law may restrict foragers from conducting their foraging activities. This will also affect the wild edible plants in that there is a likelihood that they will not be harvested. The law not only covers private lands as there are laws that restrict the access to lands that belong to the government. It is therefore important for the forager to avoid these lands so that they can avoid possible lawsuits if they are caught foraging in these lands. Though there is the freedom to roam in most of the countries, this law limits the forage on the areas they are not to forage.

Some of the areas that have been protected include forested land. Take the case of a country like Kenya, there has been the recent setting of laws which have allowed for the protection of these forests. This was done after so many activities were being conducted that have led to the depletion of many of the wild forests. As the activities were being conducted, they led to the destruction of most of the species of the edible wild plants. Most of the areas having the forests have been placed under forest reserves. These areas have limited access allowing only certain groups of people to access the land. This has limited the citizens from accessing the wild edible plants and therefore they have not been able to utilize them. These means that these wild resources will not be used.

There are some governments that allows foraging to be conducted in certain pieces of land. These lands are usually accessible for the general public and they are free to conduct their foraging activities. In the countries where there is allowed access to these lands, there are some bylaws that allows foragers to be able to conduct their foraging activities but with some restrictions surrounding them. They can conduct these activities but within the recommendations of the government so that they can do so without any trouble. Some of the bylaws have been enforced by the government or councils. With these laws in place, they can ensure that the forests as well as the wild edible plants are conserved.

The laws on the land also look into the activities conducted on the land. Most of the countries require that the activities conducted on the land to not endanger the edible wild plants. Some of the activities that have contributed to the destruction of these plants include mining. Mining involves digging up of soil so as to extract minerals or oil from under the ground. With such activities being conducted, they tamper with the wild resources. There laws that have outlined which areas can be used to conduct these activities giving forth room that allows the wild edible plants to grow and rejuvenate.

With regard to the laws concerning the land, there are those that seek to protect the nature of the soil. For the edible wild plants to grow, they need good healthy soil. This is the reason why most of the governments have prohibited the burning of vegetation in lands. The burning of vegetation leads to the destruction of the soil. It leads to the destruction of the air in the soil and also the humus in the soil will be destroyed. This will lead to most of the edible wild plants not being able to properly grow.

Most of the land laws have restricted the use of the edible wild plants amongst people. For instance, if a wild pawpaw plant is located in a restricted or private land, the foragers will not be able to access them. This means that its fruits will not be consumed by the population leading to wastage. The other land laws that have been used to control human activities being conducted on land such as mining have been put in place to ensure that the wild habitats of these plants are not destroyed. They, therefore, affect the wild edible plants positively by allowing them to survive and be used by the future generation by facilitating their management.

Among the institutional factors, there are the projects initiated by the government to conserve plant biodiversity and also the knowledge on the variety of wild edible plants. This looks into how these plants have been utilized amongst people. This has involved the governments setting aside land used for the sole purpose of the conservation of plant biodiversity. The conservation of the knowledge on the utilization of the wild edible plants involves researchers conducting studies in different areas. They get to obtain the knowledge from the source. After the knowledge is obtained, they are preserved in books or journals online allowing individuals to get access to the knowledge of these plants.

## 9.5 CULTURAL FACTORS

Cultural factors involve how the different cultural beliefs and systems will affect the utilization and management of the wild edible plants. Most of the cultural factors foster the general utilization of wild edible plants. In most of the traditional African cultures, there has been a great use of wild edible plants. This is because of the knowledge of these plants, the parts that are to be harvested and the recipes and methods used in the preparation of these plants. In the traditional setting, there was more on the dependence of these plants in cases such as drought as some of these plants are able to survive the extremely hot temperatures and are able to bear fruits.

Some of these edible wild plants were prepared for certain groups of people. For instance there some of the plant species that are prepared for pregnant women because of their high nutrient content which allows the mother to obtain the necessary nutrient that will facilitate the development of the baby. Some of these traditional foods are still being prepared for them as some of the dishes made using the wild edible plants have been used to identify the community. This affects the use of the wild edible plants by increasing the use amongst people as these plants are used in the preparation of the traditional dishes.

The cultures being upheld by certain communities to the current day and age facilitates the use of the wild edible plants. Take the case of countries such as Ethiopia, there is a need sense of culture despite the urbanization. A large number of their dishes are prepared by using the wild edible plants more so for cultural events. This affects the wild ethnobotanical plants as they are being heavily used by these cultures.

Another way in which culture has affected the wild edible plants is through the ownership of land. For most of the traditional communities, there is communal ownership of land in that the communities had the right to obtain the wild edible plant in areas that they find the plants. The communities had no restrictions on the amount of plants that could be harvested. There are some of the cultures that still own land communally and there are no restrictions on the harvesting on the wild edible plants. This affects the wild edible plants by facilitating their extensive usage amongst individuals.

People's cultures will also determine the type of wild edible plants they will consume. There are different cultures and different kinds of wild edible plants that are consumed by the communities. For a certain culture, there are certain wild edible plants that should be consumed and there are those that shouldn't. There are beliefs of the culture that may limit the individuals in the given culture not to consume these plants. There are taboos that are in line with the consumption of these plants. This is the reason why even though people are convinced that these plants are good for their health, they will not consume these plants because of the taboos related to the plant with the given plant. There are said to be some implications that will be incurred if the person consumes the plant. Some of the taboos indicate that the person may incur some diseases after consuming the wild edible plant. But this may not happen because they are just beliefs. For some of these cultures, these plants are only to be used for medicinal purposes and not for eating. This will affect how these wild edible plants are utilized.

With regard to culture, there have been some barriers that restricted people to marry within their own cultures. This was one way in which the beliefs and customs of the culture including the use of wild edible plants could be upheld. Over the years, these cultural barriers are being broken down. It can be noted by people marrying from other cultures. This has contributed to some of the barriers to the use of the wild edible plants being broken. Intercultural marriages have facilitates the taboos regarding the use of these plants to be broken hence, facilitating the significant use of these plants.

The conservation of this knowledge of these plants amongst the cultures has been one of the ways in which the knowledge of these wild edible plants has been transferred from one generation to another. For individuals who have knowledge on these plants, they have been able to acquire this information from the elder generation who obtained this information from their ancestors. They teach them on the methods used in extracting these plants, the plants that are to be harvested and the methods used in the preparation for these plants that will ensure the toxins present in these plants do not affect the individuals consuming these plants.

## 9.6 SOCIAL FACTORS

The social factors include how human beings are able to interact with one another and how their interactions may affect the environment wild edible plants included. With regard to social factors, we may look into the population pressure, urbanization and rural to urban migration among others. Over the years, there has been a significant increase in the number of populations. The growth in the population has affected the use of the wild edible plants. Over the years individuals made use of the domesticated plants as food with some of them being popular amongst people such as maize. With the ever- increasing population growth, domesticated food may not be enough to sustain the entire population. This has led to the dependence of the population on the wild edible plants. They are currently being used as a means of catering for the population. However, without any control being placed with regards to the wild edible plants it may lead to these plants being used extremely and to some extent, there may be over- harvesting that may not give the plants enough time to properly germinate and to even bear flowers or fruits. This may lead to these plant species becoming endangered.

On the aspect of population growth leading to population pressure. In as much as the increase in population has contributed to the significant use of the wild edible plants to sustain the population, there is also the challenge of the population pressure leading to the decrease in the number of these species of plants. With the challenge of the population pressure, there is a need for land to allow for the settlement of these populations. This has involved several lands that contain these wild edible plants being cleared and used to settle the ever -increasing population. This reduces the number of these wild plants. The social factors involve the interactions amongst people. This involves groups of people interacting amongst each other through a variety of activities cultural wise. The social interactions are amongst the various ways in which the knowledge of the wild edible plants were shared. Social activities such as traditional marriages are avenues in which the women were able to know what wild plants are edible and which ones are not. They are also taught on the methods used in preparing these plants.

With the social factors, there is urbanization which has significantly affected the use of the wild edible plants. Urbanization has involved the use of various technologies being used to increase the food supply and the increased domestication of plants. With urbanization, there is a strong dependence on domesticated plants. Most individuals are able to access groceries by purchasing them in supermarkets and grocery stores. With the increase in the production of the domesticated plants, there is the decreased use of the wild edible plants. Even though this is so, there is a reduced amount of these plants that are utilized amongst people in urban areas.

Some of the social factors include the level of knowledge among the people. The knowledge may include the level of education. For some schools, they teach on these plants and how they can prepare these plants. Some of the institutions of higher learning offer education on the wild edible plants. This allows them to gain information on the wild edible plants. This affects the wild edible plants in that people will be able to know which plants should be harvested and which ones should not. They are also able to identify the plant parts to be harvested. They are also able to know the methods used in preparing them.

The level of education will also affect the wild edible plants in the matter of utilization. The higher the level of education, the more the individuals become more conscious of their health. They are able to make a healthy decision with regard to the kinds of food they consume. It is generally known that wild edible plants are highly nutritious and are more of a healthy food option. The individuals will be able to make conscious decisions and opt to use wild edible plants when compared to other plants. This way they are able to gain more nutrients. The level of education will also influence how the individuals will harvest the parts of the wild edible plants. As they are more knowledgeable, they will harvest the plants with more consciousness with the understanding that these plant species need to be conserved for use by the future generation. They will also know which plant species are endangered and they will avoid them.

## 9.7 ECONOMIC FACTORS

The economic factors focus on the various areas of human living on how they can affect the wild edible plants. For instance, under the economic factors, we can look into the issue of employment. When one is employed, it means that he or she has a source of income. In most developing countries, a large percentage of the population is unemployed hence they do not have a stable source of income with some of them living from hand to mouth. This is the reason why in these areas, it is recorded that there high levels of poverty. In some of these countries, more than half of the population lives under the poverty line. In these areas, most of the population are unable to fend for themselves. They opt to use wild edible plants.

In these countries, the wild edible plants are harvested by individuals for domestic purposes and is therefore, a source of food. The fact that it is easily accessible and free is a major advantage enabling these families to have food on the table. For some of them, they harvest the wild edible plants for commercial purposes. In these developing countries, there are few laws concerning the use of wild edible plants. Therefore, individuals are not restricted to conduct commercial purposes by using the wild plants as a tool for trade. This allows them to get the money that they can use to feed and educate their children. The wild edible plants are mostly used in these countries because they are sold at a cheaper price. For individuals whose wages are not as much and they have many expenses to cater for, they opt to purchase the wild edible plants as they are cheaply sold in markets and grocery stores.

With regard to employment, as stated earlier, there are high rates of unemployment amongst individuals in developing countries. One way individuals have looked into solving the issue of unemployment is through foraging and selling of the wild edible plants. In the current day, there is an increasing demand for the wild edible plants among people. This has presented itself as a job opportunity for individuals. There is an existing market for these plants. Their job is to harvest the wild edible plants and sell them to the various stores or markets. There are those who opt to sell the plants directly to the consumers. At the end of the day, they get to earn money from selling the wild edible plants. This may affect the wild edible plants in that if there is an ever- increasing demand for these plants to a level that it cannot meet the population demand, it may be harvested to a greater extreme in that the plants will be overharvested and it may cause some of the species of these plants to become extinct.

Economic factors also look into how the various economic activities have affected the wild edible plants. For instance, activities such as agriculture has played a role both positively and negatively. Agriculture has played a positive role in the wild edible plants in that some of the ways that these plants are being conserved is through agricultural activities. For some farmers, they have picked up the practice of trying to plant these wild edible plants. For instance, if they harvest the roots from a certain plant some of them replant the roots of the plant in their gardens in the home that they will grow back. This is one of the methods used in conserving plant biodiversity. They're some of the farmers who conserve these plants by allowing them to grow in their farms.

Some of the wild edible plants are weeds a good example being the pigweed. It is occasional to observe that some farmers will uproot them and throw them away as they view them to be a nuisance and because they are known to be habitats for some of the plants in the agricultural field. There are other farmers who choose to allow these plants to grow. They have come up with measures that will reduce the habitation of pests amongst the plants. They can adopt farming techniques that will reduce the competition for water and nutrients among the plants and wild edible plants.

Agriculture has affected the wild edible plants in a negative way through the extensive cultivation that is conducted on the plant. In traditional agricultural practices, the land was cultivated and afterward was allowed to rest. This facilitated the growth of the wild edible plants more so those that are weeds. In the current day, there is continuous farming that is conducted that does not give room for the soil to rest. These activities have contributed to the loss of fertility in the soil. In as much as this affects the domesticated plants, it also affects the wild edible plants. Farming activities such as the use of fire in the clearing of land have also affected the soil and may contribute to the decreased numbers of the wild edible plants.

With the need for food to sustain the ever- growing population, it has disrupted the wild ecosystem for the wild edible plants. This is because it has led to most of the wild forests being cleared out so as to give for more room for the land to be used for agricultural practices. As most of the wild edible plants grow in these forests, the clearing of these forests has led to most of the species of these plants becoming extinct. This has led to a reduction in the amount of the wild edible plants that are available. Economic activities such as mining have also contributed to the destruction of the wild ecosystems for the wild edible plants. These activities tamper with the soil and the nutrients it contains which creates a limit in the growth of these wild edible plants.

Among the activities done in agriculture includes the application of pesticides and insecticides on the plants. These pesticides when used over time they accumulate on the soil. These toxins can be taken up by the wild edible plants that are growing in the same field making them be highly toxic. For the wild edible plants growing in the forested areas, they can also be intoxicated by these pesticides and insecticides. This is made possible in that as the chemical are being sprayed on the land, there is the excess amount that may find its way to water sources such as rivers. This is bound to happen more so in agricultural fields that are located next to river sources or next to lakes or streams. The water contaminated with these toxins find their way to these forests and are absorbed by the wild edible plants. This leads to these plants being contaminated by these chemicals.

The contamination causes the wild edible plants to be unsafe for consumption as it may lead to poisoning an in extreme cases, it might cause death. This is the reason why the government has set up methods to control the use of these pesticides. Some of the regulations require that the individual applying these pesticides and insecticides should have a license. They are also required to know the appropriate amounts of the pesticides that should be applied in the field to avoid the excessive amounts of the pesticides. They are to follow the instructions that have been laid out by the manufactures and the appropriate methods that should be used in the applying of the chemicals more so if the method involves the spraying of the chemicals. For instance, they should not spray the plants if it's windy. This will reduce the chances of the chemicals getting to water sources.

## 9.8 ENVIRONMENTAL FACTORS

The environmental factors affect the wild edible plants in the aspect of their growth. These plants are like other plants meaning that if the environment does not suit them they will not be able to survive. The environmental factors may influence the presence of these plants. The environmental factors are therefore quite significant to the wild edible plants. Depending on the species of the plant, there are conditions that suit it. Among the advantages of wild edible plants is that it allows individuals to identify the season being that there are some of these plants that grow in summer, winter, autumn and others in spring. This implies that they need a variety of environmental conditions to ensure their survival.

The environmental conditions may affect the nutrient level in the plant. This is because if the soil does not bear enough nutrients then the plant will not bear as many nutrients as plants are known to obtain these nutrients by absorbing them from the soil. Also, other factors such as the amount of water may directly affect the plant. The state where these nutrients are lacking for these plants may lead to a condition known as environmental stress which may lower the immunity of the plant making it to be under threat of pests and diseases.

In the case of the wild edible plants, some of the environmental factors may be influenced by human activities. The environmental factors influencing the growth of the wild edible plants include temperature, light, water, nutrition and humidity.

*Temperature:* Temperature is known to affect a variety of plant processes such as transpiration, germination, photosynthesis, flowering and respiration. An increase in temperature may lead to a significant increase in transpiration, respiration and photosynthesis. The temperature along with other factors may influence the transition of the plants from its leafy state to the flowering stage. Depending on the season suitable for the wild edible plant, there are temperatures that are suitable for it. There are some of these wild edible plants that flower in high temperatures while others in low temperatures.

While in the germination stage, there are of the plant species that require cool temperatures while others do well in warm seasons. This is what brings about the difference in the seasons in which the plant species germinate. The same case is experienced in the flowering of the plants. For the plants whose flowers are used as food, the temperature needs to be suitable for them to bloom. The flowering of the plants is usually a good indication of the time to harvest some of the plants. In the case of the edible wild plants whose tubers are used, the harvesting is usually recommended before the plant begins to flower as it is an indication that the plant is young and the tubers will be juicy.

The temperature will also affect the quality of the crop. Extreme temperature affects the wild edible plants by causing challenges such as impaired growth and reduced plant quality. The temperature will affect the activities of the plant such as respiration and photosynthesis. Depending on the species of the wild edible plant, there are temperatures that will facilitate activities such as photosynthesis. For some species of the wild plants, they need high temperatures to allow for the breaking of seed dormancy.

For some species of wild edible plants, their seeds are usually dormant. One of the ways in which nature takes care of the issue of seed dormancy is by the high temperature. Some of the plants whose seeds are dormant include peaches and lilies. The rate of exposure to the high temperature varies according to the species of the plant. For example, the lilies need an exposure of about six weeks.

There are plants that survive in low temperatures. Most of the plants under this category include the woody plants. For these plants, they have the ability to survive the coldest of temperatures. For these plants, they have an adaptation allowing them to survive the lowest temperatures. A sudden decrease or increase in temperature may lead to the destruction of the plants. As mentioned earlier, environmental factors have been influenced by human activities. Most of the human activities have affected and changed a variety of environmental factors. In the case of temperature, there have been activities such as industrialization.

There has been increased release of carbon dioxide in the atmosphere which has contributed to the high rates of global warming which has led to an increase in the temperature on the earth's surface. These activities have also tampered with the climates which can be demonstrated by meteorologists being unable to predict the weather. The increased rates of global warming have led to increased temperatures in different zones causing certain problems in plants such as stunted growth. It has also affected the manner in which the wild edible plants grow with some of them experiencing longer periods of seed dormancy while some have experienced decreased durations of seed dormancy.

There is the challenge that a lot of deforestation has been carried out leading to most of the areas being deserts. It has been noted that there are increased areas under deforestation. The desertification means that there has been an increase in temperature.

The flowering of the plants has also changed in that with the increased temperatures, for plants that need lower temperatures to blossom, they might not be able to flower hence there will be no production of seeds bearing these plants. The changes in temperature will alter the germination of the plants whereby if the temperatures are high, species of wild edible plants that require low temperatures will not be able to germinate. The extreme temperatures have contributed to some of the species of the wild edible plants heading to extinction. This means that the population will not be able to access the wild edible plants. There have been efforts to bring environmental problems back to normal. Some of the efforts include the reduction of activities such as deforestation. The replanting of trees has aided in maintaining environmental temperatures as plants are known to be wild environmental coolers. By so during the wild environmental temperatures will be maintained allowing the plants to grow.

*Light:* This is another factor that allows the growth of wild edible plants. In the wild edible plants like any other plants need light to allow for a number of plant processes. There are three aspects of light that aid in the growth of plants. These characteristics include quality, quantity and duration. The different aspects of light in the right quantities will facilitate the growth of the plants. By in- depth discussion of the characteristics, we get to have a better understanding of light as an environmental factor. For quantity, it is the intensity of the sunlight experienced on the plant surface. In the summer season, the light intensity is at the optimum and in the winter season, it is at the lowest. Depending on the type of plants and the season it is suitable for it, the light intensity may champion the development of the plants by facilitating photosynthesis. The optimum light intensity is needed for the given plant species of the wild edible plant to grow.

With regard to the light quality, it signifies the wavelength of the light. There are a variety of wavelengths of light in the sunlight. In the case of plants, they absorb blue and red light. This light wavelength is very important in facilitating plant growth. Both the light wavelengths are useful in the case of flowering. In the case of extreme light intensity, the plants are able to reflect the excess. As most of the edible wild plant. The light intensity also contributes to the growth of the leaves of the plant. There is also the duration of the light. This refers to the time or the period that the light is exposed to the plant surface.

The duration of the light exposure on the plant may influence the flowering of the plant. The exposure of the plant to sunlight is known to stimulate the flowering process of the plant. It also initiates responses experienced by the plant. For plants, they need uninterrupted exposure to sunlight that will ensure that the plant is able to produce the flowers. The light duration will also determine whether the plant will flower properly. According to the type of plant, one is able to classify the edible wild plants. There are those that are short-day, day-neutral or long-day. The classification will be dependent on the exposure of sunlight on the plant.

For the short-day plants, the plants can produce flowers if they have been exposed to the sun for about 12 hours which is different from the long-day plants the plants have been exposed to sunlight form more than 12 hours. The last category of plants is the day day-neutral plants, they are able to produce the flowers without and dependence on the period in which the time plants were exposed to the sun. The light aspect of plants may be tampered by human activities such as deforestation. For most of the wild edible plants more so those that grow in forests, they grow in undergrowths. Therefore, the removal of these plants may tamper with the amount of light exposed to these plants.

If the trees are cut way too much then it may lead to the exposure of the plants thereby leading to the inhibiting of the flowering process. This may lead to some of the wild edible plants not being able to produce seeds. This is the reason there are efforts being made to ensure that the forests are maintained hence there is a wild ecosystem maintained.

*Water:* Water plays an important part in the growth of plants more so for the wild edible plants. For these plants, they make use of water for a variety of processes such as photosynthesis. The different plants have their parts adapted to the environment to enable them to acquire water even though there is no rain experienced in the area. Some of the adaptations include the plants having tap roots that can grow further into the ground. This shows how water is important in ensuring the growth of these plants.

Some of the reasons why water is important are because it is crucial as it contributed to processes such as respiration and photosynthesis. It is also a crucial element in ensuring the turgor pressure in cells which contributes to the growth of the cells and also the shape of the cells. It is also the means by which the carbohydrates and minerals are transported within the plants. It also facilitates the cooling in the plants through the process of evaporation and allows for the opening and closing of the stomata. All these activities prove the importance of water for the wild edible plants.

There are a variety of sources of water that can be used by plants which include rivers, streams, lakes, rain and water in the soil. The soil water is an important factor as to when there is no rain, the plants tend to draw their water from the water present in the soil. Water affects the wild edible plants in that it should be present so as to ensure the survival of the plants. It also ensures the good health of the plant. The availability of water that can be used by the plants has been affected by human activities. Some of these activities have contributed to the decline in the amount of water. This is through the destruction of forests. Forests are known to be the water catchment areas. With the destruction of forests, it means that there will be little amounts of waters. The destruction of forests has led to desertification which has contributed too many rivers drying out. This has led to the wild edible plants not being able to access water and thereby inhibiting their growth.

*Wind:* Wind is an important factor that may affect the wild edible plants. Wind can play two roles, it can contribute to the growth of the wild edible plant or it can damage the wild edible plants. Depending on the situation, it can either have positive or negative effects on the wild edible plants. Wind can positively affect these plants in that they are an agent of seed dispersal. One of the ways in which the plants grow in different areas is if their seeds are dispersed. The wind is one of the ways in which the seeds are displaced from one area to another. This positively affects the wild edible plants by allowing them to grow in different areas. Another way in which wind is crucial for the wild edible plants is by facilitating pollination. This is applicable to the wild edible plants that are wind- pollinated. The wind is used as the agent that transports the pollen grains from one plant to another. This allows for fertilization in the plant leading to the production of fruits and seeds.

The wind can negatively affect the plant in the manner that excessive wind may lead to the destruction of the plant. For these wild edible plant species, they grow in forests as the trees have the ability to act as wind barkers reducing the strength of the wind impacting the plant. In the case where the wind is excessive, it may lead to the plant bending and in the worst -case scenario, it may lead to the stems of the plant breaking it hence the plant is destroyed. In cases of severe winds such as tornados and hurricanes, these plants may end up being totally uprooted from the soil hence the plant is destroyed. Wind can, therefore, affect the wild edible plants either positively or negatively.

*Soil nutrients:* The wild edible plants are said to be highly nutritive. Some of the nutrients they contain cannot be found in the domesticated plants. The plant nutrient content can be linked to the nutritive status of the soil in that the plants draw their nutrients from the soil. The same nutrients are transported within the plant and are retained or used by the plants. For the wild edible plants, like any other plants, they make use of the nutrients present in the soil to facilitate their growth. The soil nutrients like other factors can be affected by the activities done by man. For instance, the burning of vegetation to facilitate the clearing of land. The burning of vegetation damages soil in that the soil nutrient content is lost. This means that these nutrients may not be available for the wild edible plants leading to a reduction in plant nutrition.

Environmental factors are also referred to as ecological factors. They focus on the components of the environment contributing to the growth and the presence of these plants. They influence the type of growth that will be inhibited by the edible wild plants. It is therefore crucial that the environmental factors suit the wild edible plants as it will allow them to survive for generations to come.

## 9.9 PREDATION

Predation is under the environmental factors as the wild edible plants like other plants get preyed upon by insects. It is wild that some of these plants will be used as habitats by some insects. It is also expected that these plants will be eaten by insects. There are some insects that feed on the leaves or fruits of the wild edible plants. The same case is usually experienced with domesticated plants where other insects tend to feed on the different parts of the plant leading to most farmers opting to use a number of methods to ensure that the plants are not destroyed by the insects. With the case of the wild edible plants more so those that live in forests, there may not be the presence of chemicals or traps to capture the insects. A good example of an insect preying on these plants is the case of the silkworms feeding on the mulberry plant which may cause destruction of the berries causing them not to ripen.

This will lead to these plants being destroyed. For instance, if the insect feeds on the sap of the plant, it may lead to the plant experiencing stunted growth. Some of the implications of the insects feeding on these plants can lead to the plant dying. This will imply that there will be a limitation in the number of the wild edible plant hence there will be reduced amounts of the plant available to the general public. Another way that the insect can affect the plant is through the introduction of diseases to the plant. When the insects feed on the plants, there are wounds that are left open that become entry points for the diseases. This makes the plant to be most likely to suffer from diseases. A diseased plant signifies that the plant is not healthy and there is the possibility that it might die. This affects the wild edible plant in that it reduces their numbers and there is the challenge that these diseases may spread to the other species of wild edible plants leading to their destruction.

## 9.10 POLLINATION

Pollination is a key occurrence in a plant in that it allows for the production of seeds and fruits by the plants. The production of seeds allows the regeneration of more plants of the same species. For the wild edible plants like other plants, they can be categorized into two. They can be wind or insect -pollinated. For those that are wind- pollinated, they depend on the wind for the dispersal of the pollen grains from one plant to the other. The wind in this case is the intermediary hence, the name wind -pollinated plants. For those that are insect- pollinated, the insects are used as the means to transport the pollen grains from one plant to another.

For the insect -pollinated wild edible plants, the insects used in the pollination process are wild insects. These insects are attracted by the flowers produced by these wild plants. For these kinds of plants, their flowers are usually bright in color and they have an added adaptation in that they have the ability to produce nectar that attract these plants. Some of the wild edible plants that experience this kind of pollination are the Aloes. Some of the wild insects used in the pollination of these plants are the wild honey bees, sunbirds and butterflies among others. The presence of these insects, therefore, plays a crucial role in facilitating the fertilization in these wild edible plants.

Also among the ecological factors, there has been the need to conserve plant biodiversity. The government as well as researchers are looking for a way in which they can conserve the different plant species more so those that are heading to extinction. There are different ways that have been used in conserving this plant diversity which has involved the use of various technologies which include biotechnology. The conservation of the plant biodiversity allows for the conservation of the wild edible plant species for the generations to come more so the endangered plant species.

With regard to the environmental factors, environmental disasters have affected the wild edible plants. Some of these disasters include harsh weather climates such as drought. The presence of drought places a challenge for these plants. The water supply will be limited and the high temperatures will lead to high rates of transpiration. When the two conditions are experienced then it limits the chances of the plant's ability to survive. This leads to decreased nutrient levels in the plant and it can also cause a lack of proper development in the plants. There other wild disasters such as tornados and hurricanes have caused the destruction of these plants. They damage these plants by breaking their stems or to an extreme extent uproot the plants.

The other environmental disaster is wild fires. The wild fires destroy the habitat for the wild edible plants and the plants as well. A good example is the wild fires that have been experienced in Australia that have led to many of the wild forests being destroyed. This includes the wild edible plants being destroyed in the process. The environmental disasters negatively affect the wild edible plants.

# 10

## *Conservation of Wild Edible Plants*

## CONTENTS

## 10.1 INTRODUCTION

Wild edible plants as the name states are certain plants that are found growing in wild places meaning most of them are not usually planted by a farmer but they grow on their own. Wild edible plants are normally safe and approved for consumption by the health and food specialists. They normally grow under different conditions and they have several nutrition benefits to people who consume them.

Wild edible plants consist of a wide variety of plants. The consumption of the wild edible plants dates back to the olden times where humans used to consume wild plants and wild fruits to sustain themselves. The benefits of the wild edible plants include; it helps many people who cannot afford food to be able to escape hanger and starvation. It normally comes hand in hand to help those families with no or little source of living not to starve to death and therefore it assists many countries to escape starvation. Another benefit of wild edible plants is it helps people lost in the wilderness to find something to eat and hence gain strength to be able to find their way back. Another special use of the wild edible plants is they have found their way into farmers' gardens and hence, many people either grow them or go in search for them to gain a living from selling them to other people and therefore not only have they become a source of income but also they have become a source of food for several homes around the world. These foods also help poor communities prevent malnutrition because they offer the needed nutrients by communities that may fail to look for these nutrients due to lack of enough funds to finance them or buy them. Due to all these advantages of wild edible plants, it is of great necessity that they are conserved in the right way because lack of conserving them may lead to a great setback to community families and even countries that is if the at any time get extinct (Henry, 2010).

It is also important to note that there are two major ways of noting plants that are edible in the wild. The first way is called optimal foraging theory which says foraging organisms will balance the effort it took to search and eat that food. The other method is called the non-random plant selection.

Using these methods, a lot of wild edible plants have been discovered. They are so many that many people lack the knowledge of their existence. Among the major ones include wild onion, common burdock, common milkweed, common dandelion, lambs quarters, brambles, currants and gooseberries, blueberries and cranberries, sheep sorrel, chickweed, red clover, garlic mustard, miner's lettuce, common plantain, stinging nettle, common cattail, wild ginger, wild strawberry, American elderberry, wild rose and many others. We will be looking at the selected plants one after the other as we discuss ways they are conserved because each has a different way of conserving them and finally we look at the general ways to conserve these plants (Henry, 2010).

## 10.2 WILD ONION AND THEIR CONSERVATION

Wild onions are also known as *Allium canadense*. This plant consists of thin leaves that are thin and they grow in clumps. Many of the time they tend to sprout near flower gardens and near lawns. They grow on their own and many people have considered them as weeds and unwanted plants. Farmers get rid of them because they take them as weeds. These plants can be used as a substitute for the red onion and hence, can be used to cook meals. It has been scientifically proven and many people are using it around the world on the discovery of its advantages.

Farmers have started planting this in their farms so that they can use them in their cooking more so it has been discovered a great ingredient in the making of samosas. This, therefore, proves that they are important plants to grow and conserve because not only the root is edible but also the leaves. Therefore, among the major ways of maintaining and conserving these plants is to avoid using herbicides on them. The reason for this is that when herbicides are used in these plants, it becomes poisonous for consumption and therefore one cannot consume it anymore lest they experience food poisoning. Also, these herbicides lead to the death of the plan and therefore it goes away. So one major way of conserving them is educating farmers on the importance of this crop and encouraging them not to kill these plants with herbicides but to cultivate them appropriately so that they can grow well and help many families. Another way of preserving and conserving these crops is keeping them trimmed that will prevent their wild spread and hence, they will not be a problem to the farmers. When this crop is conserved in the right way, it will of great advantage because the farmers can sell them and they can gain income from this to buy other stuff. They can also use them as foods in their homes and will be able to save a lot of money from this. They are also a great compliment in meals and give out some needed nutrients and prevent malnutrition in homes. All this sums up to ensuring the economy of the country is stable and preventing its downfall (Henry, 2010).

## 10.3 COMMON BURDOCK AND ITS CONSERVATION

The other name for common burdock is Arctium minus berth. It has leaves that are largely simple and shaped like a heart with many branches, green stems and they develop purple flowers in July mostly. it mainly likes to sprout along roadsides and they are found in large numbers in the United States. Despite all these things, this plant is found to be harmful to be consumed by cattle because it causes the cow's milk to have a bitter taste and it entangles in the wool of sheep, therefore, spoiling the quality wool.

Despite these properties, the common burdock also has some advantages and one is that does not interfere with the crops and may fail to be a concern to many farmers unless those who keep farm animals especially, cows (Angier, 2001).

The plant has been proven by the World Health Ministry as being edible and safe. Three major parts have been proved to be okay for consumption and those include the young central stalk, the petioles and the root. However, the petioles include a lot of hard work in its preparation and, therefore, tend to consume a lot of time. Therefore, these edible plants give out nutrients that are used to prevent malnutrition and therefore these plants can be used to prevent malnutrition in kids from poor backgrounds. They are also used as a source of food because they can be eaten and therefore they enable communities to evade hunger and starvation. They can, therefore, be planted taken care of by the farmer and sold off in the market and therefore the farmer can earn some money to buy things and since most of them sprout on their own they are a great advantage to the farmers as they can easily be found hence they need to be conserved. Among the major ways of conserving them is planting them in a place where they get the appropriate requirements and watering them. Since, they are a delicacy to animals but have adverse effects, these plants must be grown in areas that are hidden and away from reach by the animals. They should be taken care of and the use of herbicides that cause effects to the plant should be avoided because just like the white onion, not only does herbicides and chemicals kill but it can also make them poisonous.

When these plants are conserved, they can enable the growth of the economy and help the countries to avoid so many problems that may arise along their way due to lack of food. Also, the country has the mandate to make sure they educate farmers on the benefits of these plants and the parts that are edible to man so that farmers may utilize this opportunity instead of them chopping off the plants and throwing them away. There is one way in which the government can also help to enable this plant not to go extinct (Angier, 2001).

## 10.4 COMMON MILKWEED AND ITS CONSERVATION

The common milkweed is also known as the Asclepias and it is mainly found in some parts of North America, Canada, and France. This plant is mostly known to grow on the sides of the road and it grows and spreads very fast. It contains properties that when eaten by insects like the butterfly they can prevent themselves from getting eaten by the birds and hence, the insects often use this mechanism to protect themselves from their predators. It can also grow in thin lawns and it can tolerate the little amount of sunlight but it is generally a sunlight plant.

The common milkweed was declared an edible plant and it was used in the past to cure diseases like lung cancer. This species is, therefore, an important crop because of the special properties that it holds and, therefore, it is very necessary to conserve it appropriately. This plant grows very fast and in clusters and it contains purplish-pink flowers. It is also very edible to insects and a lot of insects up to four hundred and fifty can feed on some part of the plant.

Due to its property of growing so fast and in clusters it should not be mixed with other crops because it may lead to the death of other crops as it can prevent them from acquiring the right minerals and water and enough sunlight. It is mainly seen as a weed and unwanted plant because of this property but in the real case, it should be planted in an area where the farmer can control its growth and a section on its own where it won't be able to prevent the growth of the other plants (Dan Dourson & Judy Dourson, 2019).

Farmers should also be educated on the benefits of the common milkweed so that they know its importance and they can get the knowledge of how they should conserve it. When farmers are educated on the benefits of the common milkweed they will cease seeing it as a weed and focus on how to conserve this crop. Thus, farmers will observe the crop and know how they can plant it and the months in which it does well and those which it does not do well and hence, more knowledge will be gained.

Since the common milkweed is a common food market for insects and a lot of insects like to feed on this plant, specific herbicides should be used so that these insects may not feed on the plant. The herbicide which is chosen should be one that will not cause any harm to humans and will not render the plant unfit for consumption, therefore, the farmers should first research to know which insecticide is suitable for the plant and is not harmful to the species. They should also cultivate the plant in the appropriate way that is put.

This plant loves sunlight and so planting them in an area where they can have enough sunlight will enable them to flourish. The holes in which they are planted in should also be like eighteen centimeters apart because they will eventually fill in the spaces as they have fast growth. This plant does not need watering unless they are planted in extremely dry conditions and also they can flourish well without any fertilizer added to them and, therefore, they are not hard plants to grow and not cause efforts to farmers and farmers can easily grow this plants to help beat starvation and common illnesses (Dan Dourson & Judy Dourson, 2019).

## 10.5 COMMON DANDELIONAND ITS CONSERVATION

The common dandelion is mainly found in areas that are disturbed by human activities. It is also found in areas with a lot of sunlight and therefore it can be found in farms and lawns mostly. The common dandelion is found in different varieties and can be found in different parts of the world. Unlike the common milkweed, the common dandelion does not grow as fast and it is, therefore, a low growing plant that requires a lot of sunlight for it to flourish and they are not easily found in the very cold places. The common dandelion is a special type of edible plant because all its parts are edible except its seeds. The common dandelion is a type of plant that has green leaves that are rugged more so-called teeth. They are used to make wine and, therefore, it is a great component. The root of the common dandelion is a taproot and it is used as a great substitute for coffee. The leaves are also used to make fruit salads of different varieties all over the world. The common dandelion is also a great source of food for bees and therefore farmers who keep bees will prefer to have this in their farms for the bees to feed on and therefore because of all these advantages the taking care of the plant is of great necessity and conserving it will of great help. Not only will it prevent hunger and starvation, but its by-products can be sold and bring income to the country family or community at large. Growing this crop is not hard and there are very few regulations that should be put in place to ensure this plant flourish as it should (Tardío, J., Pardo-de-Santayana, M., & Morales, R., 2006).

The first thing that should be maintained is taking note that the common dandelion does not grow in soils with low potassium level and therefore the farmers are advised to use fertilizers that have high levels of potassium when they are cultivating soil that these crops are supposed to be grown in. Due to the low growth of the common dandelion, the farmers should grow as many plants as possible so that they can flourish. The farmers should also be educated about the importance of the common dandelion so that they know the right ways of conserving it and taking care of it.

The common dandelion also grows in areas where there is a lot of sunlight to support its growth and therefore farmers are advised to grow them in areas where they are sure of enough sunlight being present and there will not be any shadow that will prevent sunshine from seeping through. Therefore the farmers are advised to plant in open fields for them to flourish. Lastly, the herbicides and insecticides that are used should not be poisonous to the plant and they should ensure that the plant nourishes well in the condition it is developing in (Tardío, J., Pardo-de-Santayana, M., & Morales, R., 2006).

## 10.6 LAMBS QUARTER AND ITS CONSERVATION

Lambs quarters are commonly known as the pigweed. This plant has been greatly used in most countries as they are a great source of food. The pigweed despite being considered as weeds, many people have the knowledge of pigweed being a type of food and they consume them. The pigweed is a fast -growing crop that consists of green leaves. The leaves of the pigweeds are the ones that are used for consumption. They have an amazing taste and they are prepared in the same way as the spinach or Sukuma wiki and hence a good substitute for these two.

The most common way of recognizing the pigweed is their root is a pink tap root but it has green leaves. there are different types of pigweeds and they all produce black seeds.

Pigweed is known to grow and do well in sandy soils, unlike most plants. Therefore, it is often the responsibility of the farmers to maintain these types of soil when they are growing it. Sandy soil is mostly found in dry areas and, therefore, these crops can do very well in these conditions. They do not require a lot of water for them to grow.

The advantage of this is that it enables families in dry areas to have food hence, it reduces the chances of deaths due to starvation. They give good nutrients too just like most plants of the spinach family and, therefore, they prevent lack of nutrients in poor communities as they are cheap and easily affordable. These plants have found their way in being a delicacy in many families and has found their way into the market andr a source of income to most families (Blair, 2014).

Another condition that should be maintained is the type of insecticides that are used in these crops should not cause any harm to the human race making it not good for consumption.

These crops also grow in conditions that are saline salts. They do so well in saline salts, therefore, those places which have salty water and saline salt should grow more of the lambs quarters because they are likely to flourish well in the areas as compared to the other places. The advantage of the pigweed is that they may be cultivated with other crops and they will not have any effect on the plants. They will do well together and complement each other giving each other nutrients.

Lastly just like other crops the pigweeds should be grown in areas that have adequate sunlight because they contain chlorophyll that enables photosynthesis to occur. these crops are, therefore, grown in places where there is adequate sunlight and where the sun is not hindered from seeping through to the plants. They are watered occasionally in the case of dry conditions and they are improved by using fertilizers that are recommended for their growth. Therefore, the growth of the pigweed is not a complicated process and many farmers are investing in its growth (Blair, 2014).

## 10.7 BRAMBLES AND THEIR CONSERVATION

The bramble is a large genus of the flowering plants. They are many grown by farmers due to the fruits they produce. The main advantage and edible part of the bramble are the fruits. The bramble produces fruits like the raspberries, the loganberries, the boysenberries and the blackberries. Berries have great special effects on the human body: one is they provide an increase in blood in a human body. Like all other fruits, berries are a great component of vitamins and nutrients that help in the growth of a human being. Fruits are also fulfilling and hence, they can be used as a source of meal in a case where there is no other source of food (Merritt Lyndon Fernald & Alfred Charles Kinsey, 2012).

The brambles were mostly known to grow in the wild and the fruits it produces were ranged in the category of wild fruits. Many of the times when the travelers were lost or had a long journey to cover and their food supply ran out they would most often survive on these fruits as their source of food. Since these crops were declared as edible plants as they did not have any harmful effect on the human body, human beings began to plant these crops in their homes and this saved them the pressure of having to go hunting for these fruits.

These fruits are a source of income to the farmers who plant them then they go ahead and sell them in the market and with these, they earn from it and therefore the farmers have the mandate to ensure they maintain the desired conditions for the crops to survive and give out quality produce.

Firstly, the brambles are grown in temperate regions because that is the condition it flourishes well in. Therefore, due to this, the farmers have to maintain the needed condition. Therefore, the farmers in temperate regions are most likely to plant these crops since it will flourish better in their side.

Brambles also need a lot of sunlight for it to survive, therefore, most farmers are advised to plant them in areas that have the presence of sunlight, however, they require a lot of protection from cold and wind. Therefore, farmers need to devise ways in which they can maintain this.

The wild brambles contain pests and insects which may infect the other plants that they are planted with and therefore there is a need to use the appropriate and approved chemicals in the farm so that it may not make the plant to be un suit for consumption.

The farmers are also advised to plant these crops away from other wild crops because they do not complement each other so well and there is a need to plant them in an area that is quite far from the other crops. Lastly, the farmers should be given directives by experts on how to take care of these crops and how to ensure that they are fit for human consumption. The government in the agriculture sector should educate the farmers on the importance of pigweed and how it helps the communities in different ways so that the farmers may not just treat it like a weed but like an edible plant that they can depend on for food (Merritt Lyndon Fernald & Alfred Charles Kinsey, 2012).

## 10.8 CURRANTS AND THEIR CONSERVATION

Currants are also known as dried Corinth grapes or zante currants. The currants resemble raisins but quite smaller, therefore, they look like small raisins more or less. Many people have developed a taste for them and they say that they taste better than the grapes and are often used in buns and tarts. They have a strong terrific taste in them that is very tasty and delicious and many people include them in their bakings. They are also used quite often to garnish salads because not only do they give a pleasant taste but they also make the salads look very tasty and gives them a nice color that is desirable.

These wild edible plants are very nutritious with a lot of vitamins, therefore, by growing them in their farm's, farmers can evade malnutrition in their homes and hence the families grow healthy. The price of these currants in the market is also quite good and the farmers can sell them and get some money to hold their family together. This is because a lot of bakers love baking with currants and, therefore, the currants market cannot be extinct at any one time. The other factor is since they are wild edible crops it means that they are a source of food and many people use them as a source of food and, therefore, it enables community families and nations to be able to evade starvation and hence have something to eat. Due to all these factors, the conservation of these plants is very important and many farmers must consider it (Harlan, 2011).

Just like other berries, the red currants have a short life span and therefore preserving them is key. They are preserved by storing them in the red currant jelly and when this is done they can have a longer life span. The other way of storing them is by covering them loosely and chilling them. This enables them to remain fresh for quite a longer time than when they are left on their own. The last way of preserving the currants is by drying them. Most currants in the market are sold when dry as it enables them to last longer. The advantage of drying currants is that it enables them to remain fresh for a very long amount of time and one can dry their currants by using a dehydrator.

Currant plants often thrive well cool summer regions as compared to the warm summer regions and, therefore, the farmers have to consider this when they are planting them. They should observe and see the climate of their area as this will determine if they can plant it or not. They can also carry a disease called white pine blight and hence, in those areas that white pines are grown they should control the growth of the currants in such areas (Harlan, 2011).

## 10.9 GOOSEBERRIES AND THEIR CONSERVATION

Gooseberries are also known as rash berries. They were commonly known to grow in the wild but when they were discovered as edible plants many people started planting them in their homes. The gooseberries are delicious when they are baked in pies or mixed in sweetening cream to make gooseberries fool. they are very nutritious and therefore they are recommended by many healthy people. When they are grown in the appropriate conditions, they thrive very well and with this they become ready for consumption and ready to be sold.

First, the gooseberries grow well in moist conditions. Farmers who are not in moist conditions are advised therefore to have to water their crops from time to time otherwise they will not grow as desired or they will dry up. Despite the moist conditions, the soil must be well- drained for them to thrive very well. These two conditions are key for the gooseberries to grow well. If they are maintained the result is very nice and the farmers can go ahead and sell them in the market for very good prices and, therefore, they can earn a good amount of money to maintain their families (Tull, 2013).

The soil that the gooseberries should be well -fertilized soils. Most well-fertilized soils are often found in areas where farming activities take place. If farmers are not able to find a well-fertilized place or a place with well-fertilized soil, they are advised to put fertilizers in their farms before planting the gooseberries because this is what makes the gooseberries not only sprout very fast but also maintain good quality as they grow and therefore increase the price value in the market for them to be sold.

Just like other plants, the gooseberries prefer to grow in areas where there is a lot of sunlight. They, therefore, tend to thrive more during the summer instead of during winter because in summer there is a lot of sunshine and since they prefer sunshine they thrive well at this time. Farmers are also advised that they plant the gooseberries in an open field away from shadows so that the gooseberry plants can get enough sunshine as the sunlight will go directly to them and they, therefore, make them grow well and increase the market value for the gooseberries. Lastly, the gooseberry plants are often pruned and pruning is an easy process for the gooseberries and when the plants are pruned they promote better growth and flourishing of these gooseberry plants at large (Blair, 2014).

If these conditions are maintained then the gooseberries will develop very well and hence their value increases and since gooseberries are nutritious, they offer nutrients to the communities and families and therefore they prevent malnutrition. They also help prevent or cure some diseases that may be deadly and therefore the more they have consumed the more advantage they hold for the community at large.

## 10.10 BLUEBERRIES AND CRANBERRIES AND THEIR CONSERVATION

Blueberries and cranberries like all other berries are very nutritious and they are eaten as fruits. However, the blueberries are most commonly used to make the famous blueberry pie. They are also used to make yogurts and many other delicacies. The blueberries are very sweet and they most often bring the sweet taste in any dish. Therefore, when you find blueberries in any dish be sure to experience that sweet taste that is irresistible. How these plants are grown will determine how sweet and healthy they will turn out at the end. It is therefore totally necessary that they are taken good care of so that they thrive well.

Blueberry growing activity is mostly focused in northern highbush, southern highbush and the rabbiteye types. They are blue grape-like fruits with a sweet taste. they are round in shape mostly. For blueberries to be said they are growing well in good conditions, the farmers must take note of the plant yield vigor, which means the amount of yield that they produce. A blueberry plant is known to produce a large number of blueberries and if at all the yield goes down then the farmers should do a follow up to know what might be the problem (Press, 2013).

The second thing that farmers of blueberries and cranberries must observe is the disease resistance. If the plant lacks disease resistance and begins to show signs of a disease infecting it, the farmers should improve the disease resistance of the crops. If the crops are not resistant to diseases then they may be affected and this may lead to the death of all the crops, therefore, the farmers use the appropriate and approved farm chemicals on these crops so that they can improve the disease resistance of the crops.

Another factor that must be taken into consideration by cranberry and blueberry farmers is the taste of the fruit. As stated earlier, if the fruit changes taste, it means that there is something that is not being done right that is causing this problem. The farmers are therefore advised to take note of the taste and if at all the taste changes they should carry inspection of the plants. If the taste deteriorates it means that their market will go down and therefore the income from the sales of the cranberries and the blueberries will go down and this will highly inconvenience the farmers. The farmers, therefore, are needed to observe these factors.

Lastly, the fruit storing should be maintained and it should be clean and appropriate. The benefit of maintaining the fruits appropriately is that the fruits can be stored for a long time and hence, have a longer life span. The storage must be clean because any edible thing must be stored in clean places to prevent infections spreading to the human race or diseases or even food poisoning. Therefore, if farmers observe these processes then they will be able to achieve so much from the sale of the blueberries and canberries (Press, 2013).

## 10.11 SHEEP SORREL AND ITS CONSERVATION

Sheep sorrel is a type of herb that is found in the category of flowering plants. It is a special type because it grows from March to November. Sheep sorrel has yellowish-green flowers for the male ones and reddish flowers for the female plants. It was mostly considered as a weed but since it became approved by nutritionists as an edible plant it found itself being planted by farmers in their farms as a source of food and also for sale in the market. Sheep sorrel also has red fruits and these special characteristics are the ones that make it be easily distinguished by farmers (Ballerini, 2012).

The sheep sorrel has several uses which include , it is used as a curdling agent for milk to make cheese. It is also used as green for salads and it is very nutritious and comes out so well. Another use for this is, it is a tart flavoring agent and it is also used to garnish meals and ,therefore, with all these uses, there is a great need to take care of the plants in the appropriate ways.

Among some ways of conserving the wild edible plants is planting them in acidic soils. It has been found and research has shown that the sheep sorrel does so well in acidic soils and these are the most favorable way of planting the sheep sorrel. In farming, there are ways of determining the acidity of soils and if a farmer is interested in growing the sheep sorrel he or she should first go ahead and test the acidity of their soils. If the acidity of the soil is not the required acidity for the sheep sorrel to grow in then they should go ahead and use fertilizers that have the property and ability to increase the acidity of the soil.

Another way of conserving this wild edible plant is by planting them in sandy soil. when they were first discovered they were found in sandy soils and when they were tried and tested on other soils they did not do as well as they did in the sandy soils. It is, therefore, a necessity to plant them in sandy soils for the farmers to get the results they desire for their crops and so that the sheep sorrel sprouts well (Ballerini, 2012).

A precaution to take when you plant the sheep sorrel is you should not graze your cattle near them because if cows eat them it is toxic to them. It is especially more harmful when it is consumed in a large amount by the animal. This is, therefore, another form of conserving the plants as the farmers should plant the sheep sorrel in areas that are away from their cattle so that they can prevent not only the destruction of the crops but also maintaining the health of the animals.

## 10.12 CHICKWEED AND ITS CONSERVATION

The chickweed is a green plant with green leaves and small white flowers. It is mostly found in African countries. It is a very nutritious plant which is mainly full of nutrients and vitamins. Chickweed is a great substitute for spinach just like the pigweed. It was considered a weed because of how it could fill a whole land but when it was discovered as a substitute of spinach and the fact that it was edible made a great difference and farmers begun planting them for food rather than uprooting them so that they can throw away.

The chickweed aside from being cooked, it can be used eaten raw and when it is eaten raw it is made in salads to give the salad a nutritious and fulfilling taste. It is an interesting plant considering it can be eaten when both raw and cooked. It is not hard to grow this plant and has few considerations attached to it. Farmers can also grow these plants and sell them in the market and they can use this to earn their living to support their families. They are also used as food - this kills hunger and starvation in a country and community (Sundriyal, M., & Sundriyal, D. C., 2001).

The first consideration that is attached to this crop is there is a great need to protect the plant from birds. Birds love eating this plant and if the plant is left alone, it may be devoured by the birds thus, reducing the quality and quantity of the harvest. The ways to keep the birds away include having a scarecrow on the farm to chase way these birds. You can also employ someone to chase the birds away or at least have a bird trap to trap as many birds.

Like all other green plants, the chickweed requires a lot of sunlight for germination to take place so they also grow in areas where there is enough sunlight and it is due to this reason that the plant grows in lawns side of roads and also fields. It is, therefore, the duty of the farmer who is practicing the farming of this wild edible crop to plant it where it will have access to sunlight. The farmer should avoid planting the crop below tall plants because they may hinder sunlight from seeping through to these plants.

Despite these factors, it is interesting enough that the chickweed grows in a temperate climate. It grows and sprouts very well between fifty-three degrees and sixty-eight degrees Celcius. The farmers should first observe the temperature of the area that they are growing the plant and if it is an area with adequate temperature that the plant thrives in then they have this advantage because the crop will grow well and sprout well. However, it may be a challenge due to the changing climatic conditions that are there (Sundriyal, M., & Sundriyal, D. C., 2001).

## 10.13 RED CLOVER AND ITS CONSERVATION

Red clovers are wild edible plants that have many uses in the human body. They are known to have used like they can be widely used to help prevent cancer, women use them to treat the side effects that come along with menopause, they are also used to trim weight and many other uses. The red clover is even used in foods and beverages as a spice to give the food or beverage a special and unique smell and taste. Because of all this importance, the red clover has tempted the farmers and farmers are planting it in their farms and selling them. The price of the red clover is quite high and farmers get a lot of money from this and many have been able to support their own families from this. There is, therefore, a great need for the farmers to take their time and follow the considerations that have been put up to ensure the conservation of this crop. This will be beneficial to them because the higher the quality of their products, the better it is for them as the quality and quantity of harvest will be good and hence, they will get more profit. It will also benefit the crops by preventing them from becoming extinct (Pardo-De-Santayana, M., Tardío, J., & Morales, R., 2005).

The first consideration that should be put in place is digging shallow holes for planting these crops. The holes should not be too deep because the moment they are too deep then the crops will not sprout well or even not sprout at all. The required depth is a quarter of a meter deep and this will make them grow well as desired but if this is not maintained then they will fail to grow in the way they are supposed to grow.

The other consideration that should be maintained is for red cloves to grow well is that they require fertilizers. Most farmers advocate that the fertilizers that should be used should be organic fertilizers that have potassium in them. Before you plant the seeds first you cultivate the land using this fertilizer and then you dig the holes and plant the seeds. When the crop starts growing, a fertilizer is needed until their harvesting time and with this, the crops grow strongly and the product is of good quality and also the quantity of the product is as it is desired by the farmers.

Red clovers are easy to grow because they do well in areas with an average amount of water and sunlight. Another advantage is that they can survive in areas that have or are experiencing drought and hence, this is a great thing because they can be used to prevent starvation in such places and they also help provide the needed nutrients to such people. To make the red clovers grow faster, the farmers are advised to use a fertilizer with nitrogen. Nitrogen is a nice component to help boost them and hence the red clover plants will have fast growth and it will take a short time (Pardo-De-Santayana, M., Tardío, J., & Morales, R., 2005).

## 10.14 GARLIC MUSTARD AND ITS CONSERVATION

Garlic mustard is a weed that is very hard to control once it starts growing controlling it is very hard because it grows and spreads by seed. It is a major concern in forests because it grows and spreads and even though it is edible to human beings, it is not eaten by the wild animals and therefore they are not easy to clear in the forests. Due to this, farmers are advised that they should plant in an area where they can control their growth otherwise they will develop everywhere without control and fill the whole garden.

Growing garlic mustard is not a hard process because the seeds of the garlic mustard are self-fertilizers and, therefore, they grow well on their won. The growing of the garlic onion is important because the garlic mustard is nutritious and provides the needed nutrients. It is also able to provide a source of income to the family when they sell them in the market that they can use to support their families. And hence, they must be conserved appropriately as they are required. Due to their high growth rate, the leaves must be cut off or pruning must take place to enable the managing growth of the garlic mustard plants (Pardo-De-Santayana, M., Tardío, J., & Morales, R., 2005).

## 10.15 MINER'S LETTUCE AND ITS CONSERVATION

Miner's lettuce is a green plant that has a very pleasant taste. it initially grew in the wilderness before it was discovered. It was later made an edible food by communities and it found its way into kitchen farms. Its name came about from an old experience where miners in the past used to eat this plant for them to prevent contracting scurvy. Scurvy being a disease caused by malnutrition or lacking important nutrients in the body they would look for miners lettuce and use it as their source of food hence, the name miner's.

This, therefore, shows that the miner's lettuce has specific key nutrients that are required by the human body. This means that the miner's lettuce is used by many poor communities as a source of food and, therefore, reduce malnutrition in these communities and evade starvation. Again farmers who plant the miner's lettuce can get some money from selling these plants in the market and due to their good taste, most families enjoy the miner's lettuce as a delicacy. Due to these special qualities, conserving the miner's lettuce is key in any community.

The miner's lettuce is grown in native winter and this is the time it flourishes well. Farmers must consider this when they are planting this plant to ensure that they thrive well. It also needs a lot of water and that is the main reason that it thrives well in winter because of the presence of a lot of water. Normal treatment should be offered to it and the results will be good. The miner's lettuce is a special plant and its leaves flowers and roots are edible. It is well known for its salad that is made from the leaves (Blair, 2014).

## 10.16 COMMON PLANTAIN AND ITS CONSERVATION

The common plantain is also known as the great plantain or the broadleaf plantain. It is mostly found in areas like the mid-Canada and Florida. Like most wild edible plants, the common plantain is also used as a medication to treat the common flue. It is an edible plant and has been proven as one by the health organization. It is used to treat illnesses like coughs, bronchitis and even swellings. If one is stung by a bee or insect this plant is also used to treat these stings and other sores and blisters in the body.

Among the most common conservation technique for this common plantain, is the farmer should avoid using herbicides or chemicals. A farmer is advised not to use any chemical on this plant. If at any time the farmer does not want it on his or her farm, then they should uproot it instead of spraying chemicals to them. It is a low growing crop and it grows with broad medium green leaves. To uproot the plant, you slowly and gently pull it out of the soil.

The common plantain is a remedy for most common diseases, it is highly used and many people search for it. This, therefore, means that the farmers should be educated on the benefits of this plant such that they do not consider it as a plant only or a weed but a great necessity. The research community should research ways they can help improve the growth of these plants and regions they can plant it for it to do better. As a green leaf plant, it requires a lot of sunlight for it to carry out photosynthesis and therefore this should also be a consideration to be put in place. Since it is a low growing plant, a farmer should avoid planting high growing plants with the common plantain because it may hinder the growth of these crops as it will hinder sunlight from reaching the leaves of this plant (Daniel Zohary, Maria Hopf, & Ehud Weiss, 2012).

## 10.17 STINGING NETTLE AND ITS CONSERVATION

The stinging nettle is a plant with sharp hairs on the leaves. These sharp hairs are their mechanism of protecting themselves. These hairs have chemicals such as formic acid. These chemicals irritate the skin of human beings when they touch them and may lead to reddishness on the skin of people. They are ,therefore, not safe to touch with bare hands.

These plants grow fast in any soil and have to be planted in places with adequate water. If the place lacks enough water, then the farmer is advised to water the plant just as he waters other plants. They should also be planted in well-fertilized soils so that they may thrive very well. During harvesting, the farmer should take into consideration to have protective gear such as hand gloves so that they may prevent themselves from getting stung by these plants and also so that they prevent itching and having reddish sores on their skins.

## 10.18 WILD STRAWBERRIES AND THEIR CONSERVATION

The wild strawberries are a great fruit that was known to grow in the wilderness. Just like the market sold strawberries, the wild strawberry resembles it but has a slight difference in that the wild strawberry is a little quite smaller than the other one. People consider the wild strawberries as poisonous but they have been proven as edible and safe for human consumption. The first consideration, therefore, that should be taken is educating farmers and people in communities that these plants are not poisonous but they are safe to be eaten just like the normal strawberry.

An advantage of the wild strawberries not only include the fact that they give out fruits that are used as a source of food and they are nutritious but also the fact that they grow close together and many people use them as a ground cover and other put them in vases. They are normally grown in a very easy way by just planting the seeds in empty tins or even one can use cute vases if they desire a little twist and they sprout and give out their delicious fruits after some time. The taste of the wild strawberries is not that different from the one of the normal strawberries.

The care that is required for the wild strawberries is very minimal. First of all, the wild strawberries need to be planted in areas with good water supply probably during rains so that they can thrive well because they need water to grow. During the dry seasons or the hot seasons, the farmers are advised to water these plants regularly so that they do not dry off at any one time. If this care is not taken into consideration then the leaves can wither very fast and it can lead to the death of the plant without giving out any fruit (Tardío, J., Pardo-de-Santayana, M., & Morales, R., 2006).

The second consideration that must be met is during winter the plant needs to be mulched so that the plant can be protected. This is, therefore, a great benefit because they can thrive in both the winter and summer seasons and therefore this ensures that there is a constant flow of fruits during the whole year at any time of the day. This, therefore, means families can enjoy any strawberry meal be it a pie of juice at any time and they can acquire a lot of nutrients and a lot of vitamins from this.

The last way to conserve the wild strawberry is in case of any pest infestations, the farmer has an obligation of using the desired chemicals and pesticides to kill pests so that they can protect their plant. Having the firm secured is also a nice way of protecting the plant from animals be it wild or domestic. The farmer can also use the help of a scarecrow or employ a person to chase away the birds that may come to devour on the fruits that have come out from the wild strawberries that you had planted (Tardío, J., Pardo-de-Santayana, M., & Morales, R., 2006).

## 10.19 AMERICAN ELDERBERRY AND ITS CONSERVATION

The American elderberry is a deciduous crop that is mainly found in North America. They are very easy to grow and maintain, as they are wild crops. Just like other green plants, this plant requires sunlight for it to grow and sprout

The second thing that farmers of the American elderberry must observe is the disease resistance. If the plant lacks disease resistance and begins to show signs of a disease infecting it, the farmers should improve the disease resistance of the crops. If the crops are not resistant to diseases then they may be affected and this may lead to the death of all the crops, therefore, the farmers use the appropriate and approved farm chemicals on these crops so that they can improve the disease resistance of the crops.

Another factor that must be taken into consideration by American elderberry farmers is the taste of the fruit. As stated earlier, if the fruit changes taste it means that there is something that is not being done right that is causing this problem. The farmers are advised to take note of the taste and if at all the taste changes they should carry inspection of the plants. If the taste deteriorates it means that their market will go down and, therefore, the income from the sales of the American elderberry will go down and this will highly inconvenience the farmers. The farmers, therefore, are needed to observe these factors (Tardío, J., Pardo-de-Santayana, M., & Morales, R., 2006).

When there is growing of the American elderberry, the soils need to fertile so the farmers are advised to use the correct and appropriate fertilizers. Lastly, the fruit storing should be maintained and it should be clean and appropriate. The benefit of maintaining the fruits appropriately is that the fruits can take a long time and hence have a longer life span. The storage must be clean because any edible thing must be stored in clean places to prevent infections spreading to the human race or diseases or even food poisoning. Therefore, if farmers observe these processes then they will be able to achieve so much from the sale of the American elderberry (Tardío, J., Pardo-de-Santayana, M., & Morales, R., 2006).

## 10.20 WILD ROSE AND ITS CONSERVATION

The wild rose is a special type of wild edible plant that has white pinkish or yellowish flowers. The wild rose is a large genus of the flowering plants.

They are many grown by farmers due to the fruits they produce. The main advantage and edible part of the wild rose are the fruits. Like all other fruits, they are a great component of vitamins and nutrients that help in the growth of a human being. Fruits are also fulfilling and hence they can be used as a source of meal in a case where there is no other source of food (Dan Dourson & Judy Dourson, 2019).

The Wild rose was mostly known to grow in the wild and the fruits it produces were ranged in the category of wild fruits. The wild rose has reddish small fruits that are found beneath the flowres.

Many of the times when the travelers were lost or had a long journey to cover and their food supply ran out they would most often survive on these world fruits as their source of food. Since these crops were declared as edible plants as they did not have any harmful effect on the human body, human beings began to plant these crops in their homes and this saved them the pressure of having to go hunting for these fruits.

These fruits are a source of income to the farmers who plant them then they go ahead and sell them in the market and with these, they earn from it and therefore the farmers have the mandate to ensure they maintain the desired conditions for the crops to survive and give out quality produce.

Firstly, wild rose is grown in temperate regions because that is the condition it flourishes well in. They grow without the help of anyone and they nourish on their own. This is an advantage because farmers will not have so much work caring for them (Dan Dourson & Judy Dourson, 2019).

Just like all other green crops the wild rose also needs a lot of sunlight for it to survive this means that the farmers are advised to plant them in areas that have the presence of sunlight.

In case any of the white rose contains pests and insects there is a need to use the appropriate and approved chemicals in the farm so that it may not make the plant be unsuitable for consumption and so that they can kill these insects that may prevent the appropriate growth.

The wild rose has an advantage since it grows in any soil and, therefore, it can be grown anywhere. It is a tough breed and, therefore, it can grow on its own without the help of the farmers. However, there is one breed of this plant that is known to grow in wet soils. The flowers produces sprouts and takes the plant into winter.

Lastly, the farmers should be given directives by the experts on how to take care of these crops and how to ensure that they are fit for human consumption. The government in the agriculture sector should educate the farmers on the importance of this plant and how it helps the communities in different ways so that the farmers may not just treat it like a weed but like an edible plant that they can depend on for food (Tull, 2013).

## 10.21 ADVANTAGES OF CONSERVING WILD EDIBLE PLANTS

The first advantage of wild edible plants is that they are used as food. When these plants are eaten, they help the people to gain strength and hence, prevent starvation. Since the crops are wild crops, they grow on their own and, therefore, they are not very expensive in the market and therefore many people can buy them or afford to buy them if they are not farmers or they cannot plant their own.

The second advantage is the farmers can sell these plants in the market and make a living from them. Many people have sought to planting these crops in their farms since they do not have a lot of requirements for growth and the likelihood of experiencing a loss is quite low, therefore, farmers prefer to plant and sell these plants. When they sell them, they gain an income that they can use to sustain their families or even to support another business. Therefore, the conservation and planting of wild edible plants is a method of earning income for most farmers (Tull, 2013).

The third advantage of the conservation of wild edible plants is that they help hinder the spread of diseases. Most wild edible plants are used as medication. Some are boiled to produce the healing factor, others are eaten raw and even other applied directly on the skin. In the olden days before medication was invented and drugs came about, the traditional healer used to use these techniques in curing several diseases such as the colds. Some of these plants are even stronger than modern drugs. This, therefore, shows that there is a great need for individuals to conserve these plants as they help prevent and cure diseases.

The fourth benefit of wild edible plants is that they help families and communities to prevent malnutrition. As we have seen most of these crops have nutrients in them that are necessary for the normal growth if the human body. Human beings are therefore advised to take these crops so that they may prevent malnutrition and so that they may acquire a lot of nutrients in their body for strong and healthy bodies. Wild edible plants are also used to make medicine that is used in the modern world. If you read past the ingredients of modern drugs you can be sure to find one being the wild edible plant that is why there is great need to conserve them (Tull, 2013).

Lastly, the wild edible plants are used in the kitchen for adding flavors in meals. Most wild edible plants have a good taste attached to them and when they are used in kitchen meals they complement the dish giving the dish a unique and delicious taste. An example is the blueberries. When the blueberries are added to cakes and pies they make the cake and pie to be extra sweet but in a good way. This, therefore, complements the dish.

This overall shows that the preservation and conservation of wild plants is an important part because it has so many benefits and it can help a country in a time of crisis to evade things like hunger starvation and poverty (Auerbach, 2011).

# *References*

1. Angier, B. (2001). Feasting Free on Wild Edibles. Pennsylvania: Stackpole Books, 17-112.
2. Auerbach, P. S. (2011). Wilderness Medicine E-Book: Expert Consult Premium Edition - Enhanced Online Features. Netherlands: Elsevier Health Sciences, pp2304.
3. Ballerini, L. (2012). A Feast of Weeds: A Literary Guide to Foraging and Cooking Wild Edible Plants. California: Univ of California Press, 1-91.
4. Blair, K. (2014). The Wild Wisdom of Weeds: 13 Essential Plants for Human Survival. Hartford: Chelsea Green Publishing, 1-73.
5. Dan Dourson & Judy Dourson. (2019). Wild Yet Tasty: A Guide to Edible Plants of Eastern Kentucky. Kentucky,: University Press of Kentucky, 1-36.
6. Daniel Zohary, Maria Hopf, & Ehud Weiss. (2012). Domestication of Plants in the Old World: The Origin and Spread of Domesticated Plants in Southwest Asia, Europe, and the Mediterranean Basin. London: OUP Oxford, 175.
7. Department of the Army. (2009). The Complete Guide to Edible Wild Plants. New York: Skyhorse Publishing, Inc, pp160.
8. Harlan, J. R. (2011). Origins of African Plant Domestication. Berlin: Walter de Gruyter, 3-67.
9. Henry, R. J. (2010). Plant Resources for Food, Fuel and Conservation. London: Earthscan, 7-91.
10. Jman Redzic, S. (2006). Wild edible plants and their traditional use in the human nutrition in Bosnia-Herzegovina. . Ecology of Food and Nutrition, , 45(3), 189-232.

11. Ladio, A. H., & Lozada, M. . (2004). Patterns of use and knowledge of wild edible plants in distinct ecological environments: a case study of a Mapuche community from northwestern Patagonia. . Biodiversity & Conservation, , 13(6), 1153-1173.

12. Lyle, K. L. (2010). Complete Guide to Edible Wild Plants, Mushrooms, Fruits, and Nuts: How to Find, Identify, and Cook Them. Maryland: Rowman & Littlefield, pp224.

13. Lytton John Musselman & Harold J. Wiggins. (2013). The Quick Guide to Wild Edible Plants: Easy to Pick, Easy to Prepare. Maryland: JHU Press, 1-97.

14. María de Cortes Sánchez-Mata, & Javier Tardío. (2016). Mediterranean Wild Edible Plants: Ethnobotany and Food Composition Tables. Berlin: Springer, pp478.

15. Merritt Lyndon Fernald & Alfred Charles Kinsey. (2012). Edible Wild Plants of Eastern North America. Massachusetts, United States: Courier Corporation, pp480.

16. Nyerges, C. (2016). Foraging Wild Edible Plants of North America: More than 150 Delicious Recipes Using Nature's Edibles. Maryland: Rowman & Littlefield, 7-91.

17. Pardo-De-Santayana, M., Tardío, J., & Morales, R. (2005). The gathering and consumption of wild edible plants in the Campoo (Cantabria, Spain). International Journal of Food Sciences and Nutrition, , 56(7), 529-542.

18. Press, A. (2013). Edible Wild Plants for Beginners: The Essential Edible Plants and Recipes to Get Started. California: Callisto Media Inc, pp258.

19. Richard J. Medve & Mary Lee Medve. (2010). Edible Wild Plants of Pennsylvania and Neighboring States. Pennsylvania: Penn State Press, 3-17.

20. Saunders, C. F. (2012). Edible and Useful Wild Plants of the United States and Canada. Massachusetts, United States: Courier Corporation, pp336.

21. Stein, M. (2008). When Technology Fails: A Manual for Self-Reliance, Sustainability, and Surviving the Long Emergency, 2nd Edition. Hartford: Chelsea Green Publishing, pp493.

22. Sundriyal, M., & Sundriyal, D. C. (2001). Wild edible plants of the Sikkim Himalaya: Nutritive values of selected species. Economic Botany, 55(3), 377.

23. Tardío, J., Pardo-de-Santayana, M., & Morales, R. (2006). Ethnobotanical review of wild edible plants in Spain. Botanical Journal of the Linnean Society, , 152(1), 27-71.

24. Tull, D. (2013). Edible and Useful Plants of the Southwest: Texas, New Mexico, and Arizona. Texas: University of Texas Press, 1-72.

# INDEX